Copper, Brass, and Bronze Surfaces

ZAHNER'S ARCHITECTURAL METALS SERIES

Zahner's Architectural Metals Series offers in-depth coverage of metals used in architecture and art today. Metals in architecture are selected for their durability, strength, and resistance to weather. The metals covered in this series are used extensively in the built environments that make up our world and also attract and fascinate the artist. These heavily illustrated guides offer comprehensive coverage of how each metal is used in creating surfaces for building exteriors, interiors, and art sculpture. The series provides architects, metal fabricators and developers, design professionals, and students of architecture and design with a logical framework for the selection and use of metal building and design materials. Forthcoming books in Zahner's Architectural Metals Series will cover steel and zinc surfaces.

Titles in Zahner's Architectural Metals Series include:

Stainless Steel Surfaces: A Guide to Alloys, Finishes, Fabrication, and Maintenance in Architecture and Art

Aluminum Surfaces: A Guide to Alloys, Finishes, Fabrication, and Maintenance in Architecture and Art

Copper, Brass, and Bronze Surfaces: A Guide to Alloys, Finishes, Fabrication, and Maintenance in Architecture and Art

Copper, Brass, and Bronze Surfaces

A Guide to Alloys, Finishes, Fabrication, and Maintenance in Architecture and Art

L. William Zahner

WILEY

Cover image: © wepix / Getty Images
Cover design: Wiley

This book is printed on acid-free paper.

Published by John Wiley & Sons, Inc., Hoboken, New Jersey.
Published simultaneously in Canada.

Library of Congress Cataloging-in-Publication Data:

Names: Zahner, L. William, author.
Title: Copper, brass, and bronze surfaces : a guide to alloys, finishes, fabrication, and maintenance in architecture and art / L. William Zahner.
Description: Hoboken, New Jersey : Wiley, 2020. | Series: Zahner's architectural metals series | Includes bibliographical references and index.
Identifiers: LCCN 2019045098 (print) | LCCN 2019045099 (ebook) | ISBN 9781119541660 (paperback) | ISBN 9781119541677 (adobe pdf) | ISBN 9781119541684 (epub)
Subjects: LCSH: Copper. | Brass. | Bronze. | Architectural metal—work. | Art metal—work.
Classification: LCC TS620 .Z34 2020 (print) | LCC TS620 (ebook) | DDC 739—dc23
LC record available at https://lccn.loc.gov/2019045098
LC ebook record available at https://lccn.loc.gov/2019045099

10 9 8 7 6 5 4 3 2 1

This book is in honor of Salvatore Orlando.
He was a good friend and advocate of the red metal.

Contents

Preface

The passage of time is reflected in the color of copper.

Of all the metals used in art and architecture, copper is the most engaging.

Throughout history, mankind has had a special relationship with copper. Copper has a weight and feeling of substance. It can be shaped and formed into useful objects, and more importantly, it has an appearance as natural as the colors of an oak forest in the fall: a color that shows value and the passage of time.

Power and force are needed to shape the other metals used in art and architecture. Copper, on the other hand, shapes and moves under the blows of handheld chasing tools. It can be easily folded, curved, stretched, and embossed. One gets close to the metal when working with copper.

Throughout human time copper and copper alloys have played different and expanding roles. Bronze sculptures of ancient deities and heroes have outlived their civilizations, even while resting under the sea for thousands of years. Copper has been mined by every major civilization and converted and cast into both useful tools and decorative statues.

The maritime world embraced copper alloys, particularly brass. Alloys with names such as Admiralty Bronze and Naval Brass recall a time when this metal served as biocladding on the underside of a ship and ornamentation on the top. Consider that the military term "brass" relates to someone of high rank: the one with the brass metals or the brass-adorned hat.

Copper alloys, both those with new, untarnished surfaces and those with colorful patinas, offer the artist and the designer an amazing palette of color to choose from and design with. Oxide colors will be predictable to a point, but beyond that it is nature that will take over the design. These colors will also act as potent inhibitors of corrosion. But note that both natural, untarnished surfaces and beautiful patinas on art and architecture forms will require something additional—either in the form of a coating or in the form of energy applied from an elbow—to keep them looking good.

I have worked with copper and many of its alloys for decades. I have hammered it, cut it, welded it, and shaped it into beautiful pieces of art. I have experimented with creating color on the surface of the metal with chemical interaction, heat, and selective electroplating. I have worked with friends to cast it and I have formed its sheets to create incredible surfaces.

Copper is special for its amazing ability to be shaped and stretched and for its ability to react with other elements and compounds to achieve a unique and beautiful surface.

Copper can be cast into glass, and the glass accepts it. It can be severely shaped, and it yields to take the new shape. It can withstand the attack of powerful acids and bases, all the while forming a natural mineral surface that slows further reaction while giving a beautiful patina.

This is the third in a series of books on metals used in art and architecture. I have attempted to cover many of the copper alloys that have found their way into use in art and architecture. New and innovative uses of the metal, along with advances in fabrication techniques and a renewed interest in patination, are fueling a renaissance in the use of copper alloys.

It is the surface of copper alloys—the part that interacts with the environment and the part that absorbs and reflects light in unique and special ways—that we find interesting as a material of design. But industry is also looking for ways of capitalizing on the antimicrobial benefits of the copper alloy surface. Tests have proven bacteria and viruses do not thrive and will diminish when in proximity to copper ions.

The corrosion resistant behavior of the copper alloy surface is unmatched. Few corrosive materials can get through the protective behavior of the oxide surface. All of this is an important and essential quality of the metal, but it is the unique color that draws us to it. There is no other metal that comes close to the intense beauty of copper alloys.

The early alchemists associated each of seven metals with a planet. Copper, one of the oldest metals known to man, was associated with the planet Venus, one of ancient man's "wandering stars." Copper was thought to represent the characteristics of feminine beauty, caring and nurturing, love and lust.

The symbol for copper is the female symbol that is also the symbol for the planet Venus.

The book is intended to give the artist and designer more knowledge about copper, its alloys and this amazing surface. Those who are interested in the metal will acquire information on how to work with the different copper alloys, how they will interact with the environment with time and exposure. Copper and copper alloys have a vast history, but the story is far from over.

CHAPTER 1

Introduction—Element 29

Copper and Copper Alloys

Cu—Cuprum

INTRODUCTION

Of the metals, only copper and gold possess colors other than gray or silver in their natural forms. When copper is combined with different elements, elegant colors and tones can be produced both naturally and artificially. This is the reason why our ancient ancestors first worked with this metal: they could identify it among rocks and stones more easily and it was more abundant than gold. The brightly colored ores of copper, malachite, and azurite surely attracted the attention of early humans. These were not your normal rocks.

Copper is element 29 in the periodic table (Figure 1.1). It falls between element 28 (nickel) and element 30 (zinc). Copper is in the same group as silver and gold: metals that it mixes with and that possess similar properties of electrical and thermal conductivity. Like copper, gold and silver were also highly valued by early man. Being more abundant, copper took on the heaviest workload. The age named for it marks an advance in civilization.

Copper possesses a face-centered cubic structure in its pure state, but this structure changes as alloying elements are added (Figure 1.2). This face-centered structure is shared with many other metals, such as aluminum and iron.

The atomic makeup of copper is responsible for many of the unique attributes this metal offers (Figure 1.3). The atomic number of 29 means that copper has 29 protons in the nucleus with 29 electrons making up its outer shells. It is the lone electron in its fourth orbital that gives copper one of

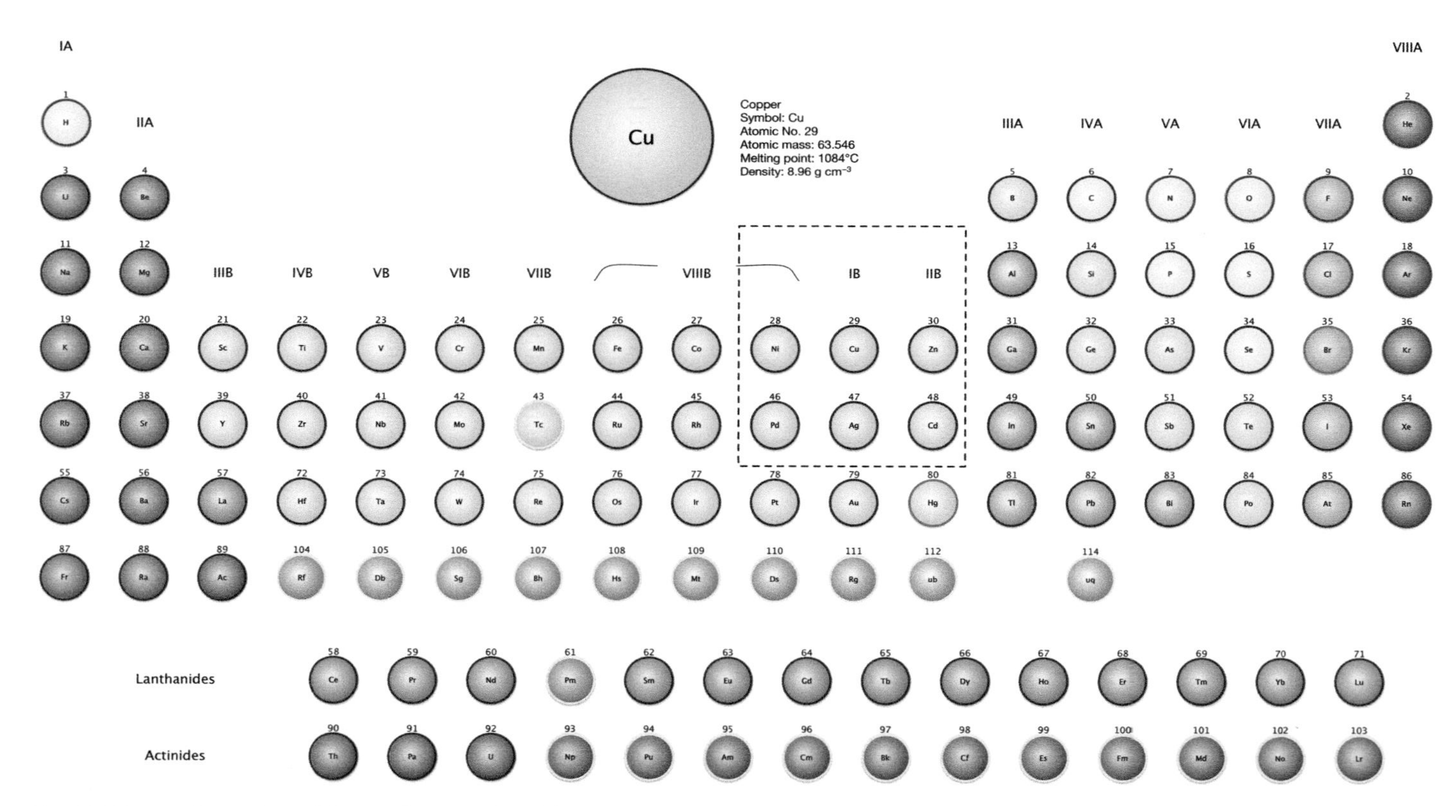

FIGURE 1.1 Copper's position in the periodic table of elements.

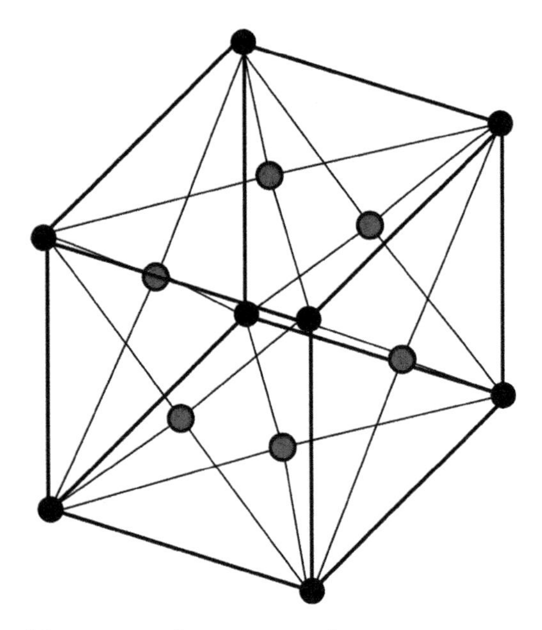

FIGURE 1.2 Face-centered cubic structure of a copper crystal.

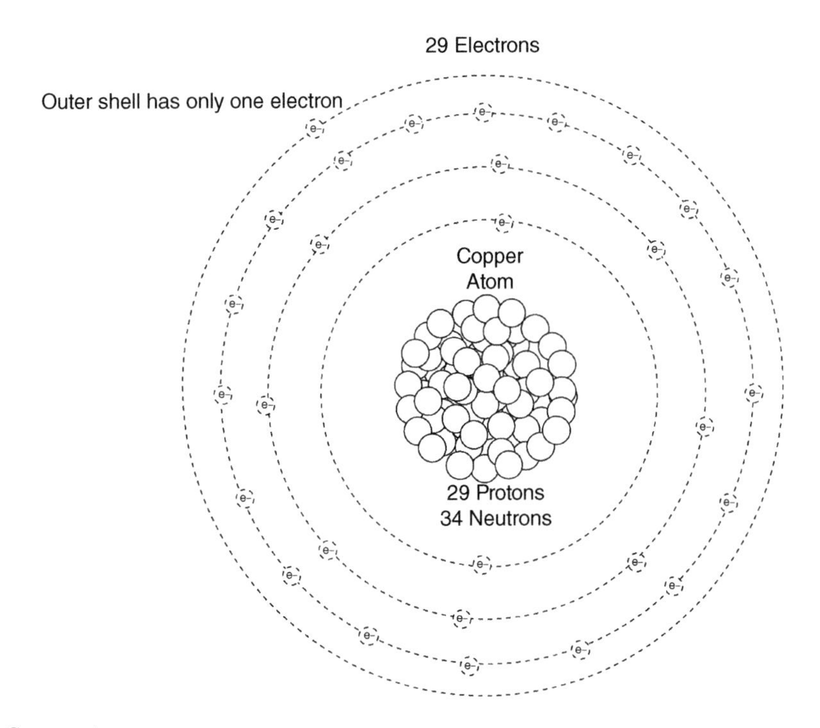

FIGURE 1.3 Copper atom.

its most important traits. This electron is free to move about, allowing electrical current to move easily from atom to atom.

This lower resistance coming from a single electron in the outer orbit gives copper its exceptional ability to transport electricity. The metals whose atoms have the fewest electrons in their outer (or valence) shell offer the least resistance to the movement of electricity from one atom to the next. Gold, silver, and copper are metals that have only one electron in their valence shells, which gives them their ability to conduct electrical current more efficiently than atoms with more than one electron. This lone electron moves freely around and through the lattice structure of the metal, transferring the electrical current with little resistance. Aluminum has three electrons in its valence orbit, while zinc, nickel, iron, and titanium have two (Table 1.1).

Operating under a principle similar to electrical conductivity, the conductivity of heat has the same order across metals. Alloying will change the heat conductivity of a particular metal, as it does the electrical conductivity. For instance, most bronze alloys of copper will not conduct as well as many aluminum alloys. The alloying elements in bronze diminish its ability to move an electrical charge.

As they are essentially shared, these valence electrons are free to flow in and around the atoms, creating a "sea" of charged particles. This sea of electrons allows the charge to move rapidly and with little resistance as energy is transferred from electron to electron that collectively make up the sea around the copper atoms (Figure 1.4).

Most of the copper found on the Earth's surface is from hydrothermal activity that brought the metal to or near the surface. Other surface copper is drift copper, deposited by glacier activity and randomly set in rubble. Copper makes up approximately 0.0068% of the upper mineral crust of the Earth and is widely distributed with concentrations in select regions.

Copper has a poor strength-to-weight ratio as compared to other metals used in industry. However, copper alloys, such as brass alloys, have a strength-to-weight ratio equivalent to stainless steels.

TABLE 1.1 Electrical conductivity of various metals in siemens m^{-1} at 20 °C.

Metal	Siemens m^{-1}
Silver	6.30×10^7
Copper	5.98×10^7
Gold	4.52×10^7
Aluminum	3.50×10^7
Zinc	1.68×10^7
Nickel	1.43×10^7
Iron	1.04×10^7
Titanium	1.80×10^6

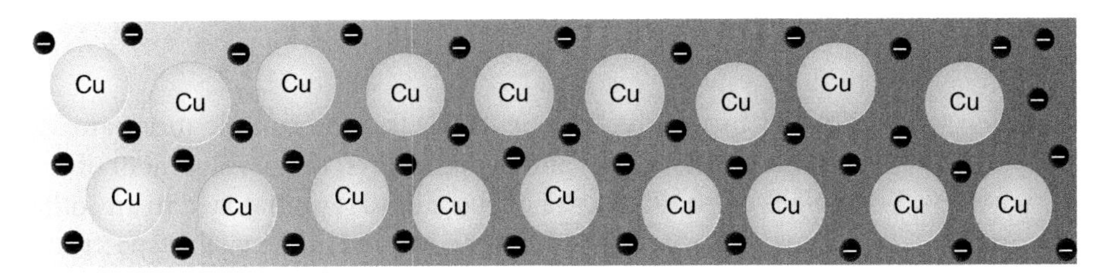

FIGURE 1.4 Sea of shared electrons around the metal atoms of copper.

Other characteristics of copper include excellent ductility, a deep forming ability, high fracture toughness, high elasticity (resiliency under shock loading), and soft edges.

Copper is also nontoxic—although copper salts are considered ecotoxic in certain instances—and has superior corrosion resistance in many natural environments.

Copper: Element 29

Atomic number 29	
Crystal structure	Face-centered cube
Main mineral source	Chalcopyrite and chalcocite
Color	Salmon red
Oxide	Brown to black
Density	8960 kg/m^{-3}
Specific gravity	8.8
Melting point	1083 °C
Thermal conductivity	401 W/m^{-1} °C
Coefficient of linear expansion	16×10^{-6}/°C
Electrical conductivity	100% IACS
Modulus of elasticity	110 GPa
Finishes	Mill specular and nonspecular Polished satin and mirror Glass bead

Copper can be painted but this is rarely done.
Porcelain enamel is an art process used extensively on copper.
Plating with other metals such as silver, nickel, and gold is easily accomplished on copper

(continued)

(continued)

Artificial patina	Greens, browns, yellows, reds, blacks, and combinations are achievable on copper and copper alloys; the development of chemical patinas on the surface of copper alloys is unmatched in any other metal
Bright appearance	Copper absorbs and reflects the red end of the visible spectrum; alloys alter this reflection by emitting yellow wavelengths along with the red end of the visible spectrum
Reflectance	
of ultraviolet	Very good
of infrared	Poor; copper absorbs infrared wavelengths
Relative cost	Medium
Strengthening	Cold working is the main method used to strengthen copper and copper alloys
Recyclability	Very easily recycled; recycled copper and copper alloys retain a high value
Welding and joining	Copper and copper alloys can be welded, brazed, and soldered
Casting	Copper and copper alloys are frequently cast in all casting methodologies
Plating	Copper and copper alloys can be electroplated
Etching and milling	Copper and copper alloys can be etched and chemically milled

IACS = International Annealed Copper Standard; GPa = gigapascal.

COLOR

Another attribute of copper and its alloys is color. There are only two metals that possess a tone other than gray or silver: gold and copper.

When light falls on the metal surface it is intensely absorbed by the atoms at the surface. The electromagnetic wave that we call light only penetrates a very small fractional distance—less than a wavelength—into the metal's surface. But this absorption is intense due to atomic characteristics specific to metals and the electrons that make up the sea that flow around the atoms.

This intense absorption of the electromagnetic wave on the surface causes a pulse of alternating current, which then excites this sea of charged particles and reemits light. This is luster: the intense reflection from a polished metal surface. The smoother the surface is, the greater the reflection. If the surface is coarse, a diffuse reflection occurs. For example, in the case of tarnish—a thickened, diffuse oxide that can develop on the surface of copper—a contrasting darker surface is apparent, dulling the copper's luster. Tarnish is a mineral formation that captures the electrons and makes the metal slightly less conductive.

The color of copper and gold is determined by the makeup of their atoms. When a light wave strikes a copper or gold surface, the portion of the wavelength from 600 to 700 nm is strongly absorbed, as Figure 1.3 shows. In metals, this absorption leads to reemission as reflected light. At the same time, both of these metals absorb the wavelengths at the blue and violet end of the spectrum poorly. This gives copper a reddish color and gold a yellowish color. It is this significant drop-off in absorption—strong on one end of the spectrum and weak on the other end—that gives these metals their characteristic color.

For example, iron absorbs the wavelengths associated with blue and violet much more than copper, but it does not absorb the wavelengths associated with red, orange, and yellow. It has a fairly flat absorption line across all the wavelengths.[1] Stainless steel is similar to iron's reflectivity, but the added chromium increases the reflection of portions of the light wave over 60%. Stainless steel is said to reflect, on average, 60% of the light wave (Figure 1.5).

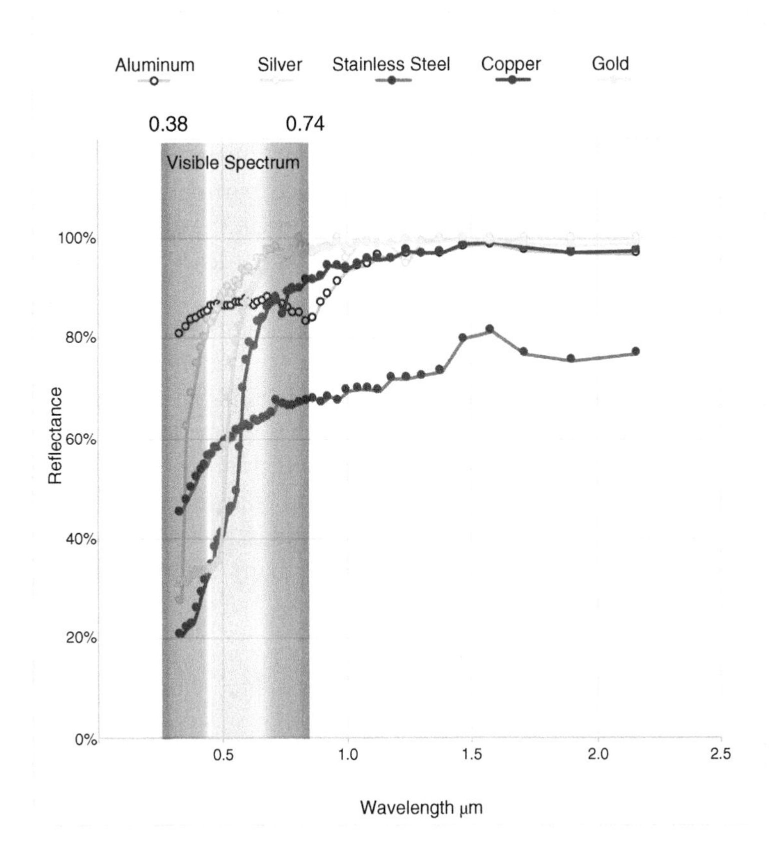

FIGURE 1.5 Reflectivity of aluminum, silver, stainless steel, copper, and gold.
Source: Data plotted from *NASA Technical Note D*-5353, "Solar Absorptances and Spectral Reflectances of 12 Metals for Temperatures Ranging from 300 to 500 K."

[1] Kurt Nassau, *The Physics and Chemistry of Color* (New York: Wiley, 1983), 161.

COLORS OF ALLOYS

The color of copper alloys is determined by several factors. The addition of various alloying constituents influences the color up to the point at which the crystal structure of the metal changes. When this occurs, the density and form of the crystal changes and light absorption can potentially change as well.

For example, as zinc is added to molten copper it dissolves and integrates into a single crystal structure, or phase, along with the copper atoms. The alloy becomes progressively more yellow in color. At the point at which 35% zinc is alloyed with copper, the metal reaches a saturation point and the color will be yellow. This is the alloy C27000, also known as yellow brass. As zinc is added beyond this 35% level, a phase change occurs in the crystal makeup, the yellow color loses intensity, and the color goes back to a bronze tone. The copper crystal lattice can no longer take the zinc atoms in and two phases develop: an alpha phase, with a face-centered cubic structure, and a beta phase, with a body-centered structure. For example, alloy C28000, commonly known as Muntz metal, contains 40% zinc and is less yellow. Both alpha and beta grains are apparent in alloy C28000 (see Figure 1.6 for comparisons of the natural colors of copper alloys).

Mechanical characteristics also change as alloying constituents are added. As zinc is added, brass alloys get stronger. However, their corrosion resistance, particularly to the condition known as dezincification, will decrease. Once the dual phase appears in the alloy, which occurs at around 40% zinc, cold working ability declines. The C28000 alloy is harder to cold work than alloys with less zinc.

Copper and its alloys have the ability to create amazing colors when they combine with nonmetal elements such as sulfur, chlorine, carbon, and oxygen (Figure 1.7). Almost everyone is familiar with the beautiful patinas that are apparent on copper roofs built a century ago. These attractive, natural-looking green surfaces developed over time and with exposure to the atmosphere. They were not precolored but allowed to absorb the carbon, sulfur, and chlorine from the air. In a real sense, copper captures the industrial pollutants of sulfur and carbon dioxide from the air and forms these beautiful surfaces composed of copper sulfate, copper chlorides, and copper carbonates.

For instance, the green patina adorning the roofs of many centuries-old buildings in cities around the world is a form of the mineral brochantite. This mineral has the formula $Cu_4SO_4(OH)_6$

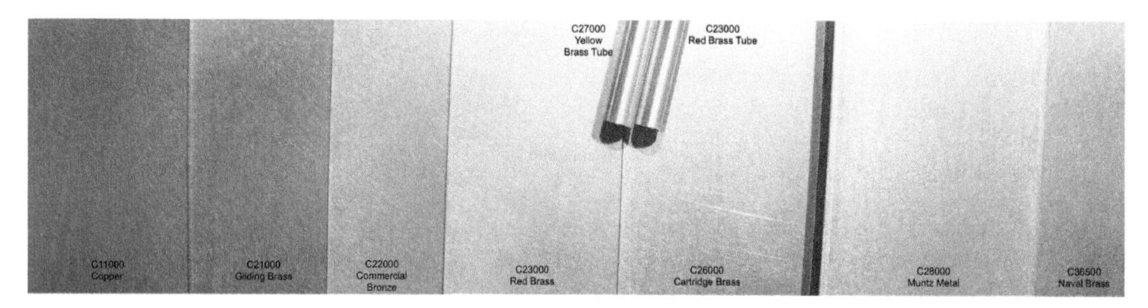

FIGURE 1.6 Image of different alloys and colors.

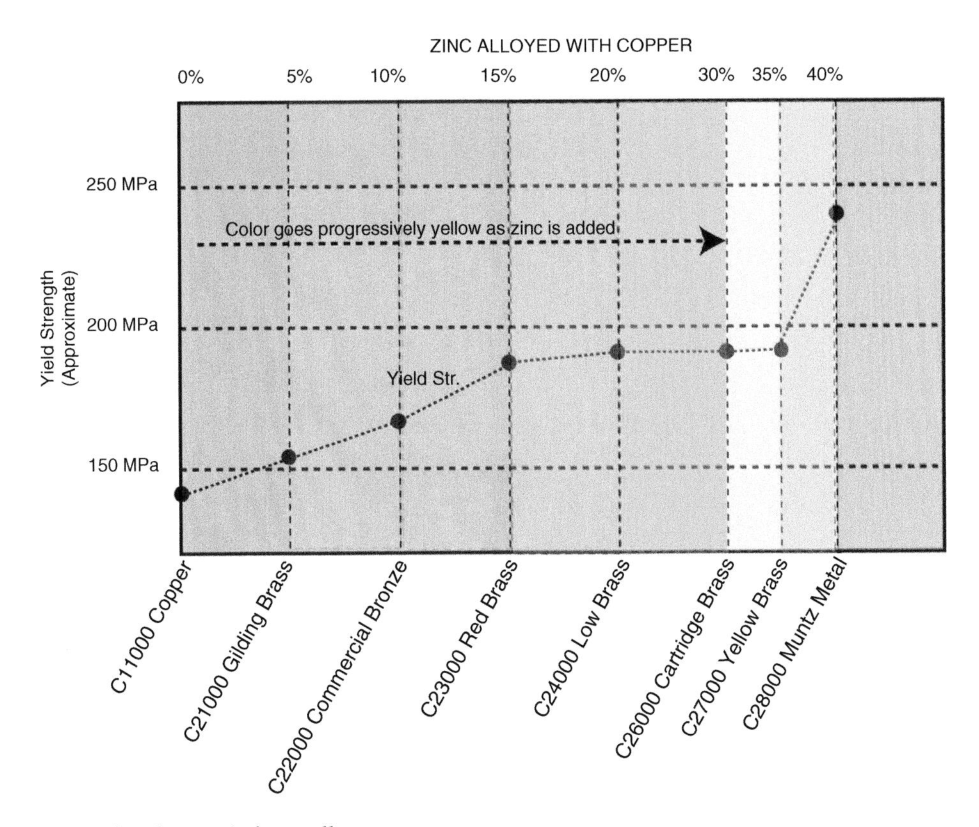

FIGURE 1.7 Color changes in brass alloys.

and displays the characteristic color of pale green. The copper began as a bright salmon-red color, but as it was exposed to the air, natural humidity, and rain the copper combined with the sulfur and formed the mineral on the surface (Figure 1.8).

The patina is the copper surface corroding, but the difference is that once formed, this tightly adhered compound protects the underlying metal. The rate of corrosion slows way down as this inert, mineral form of copper achieves a level of equilibrium with the surrounding environment.

The Statue of Liberty (Figure 1.9) initially had the color of a penny when delivered. It was not polished copper but had the rich copper color of a slightly aged penny. Dedicated in 1886, it was subject to years of exposure to the polluted environment of industrial New York in the later part of the 1800s and early 1900s. Exposed to both the chloride-rich seaside and pollution from heavy industry on the East Coast, the green patina we see today formed. As the inert mineral layer of brochantite, antlerite, and atacamite that made up the patina formed, it protected the copper plates from degradation.[2]

[2]Thomas Graedel and John Franey, *Formation and Characteristics of the Statue of Liberty Patina* (Houston: National Association of Corrosion Engineers, 1990), 101–108.

FIGURE 1.8 A naturally formed patina.

FIGURE 1.9 The Statue of Liberty.

FIGURE 1.10 *Perseus with the Head of Medusa*, by Benvenuto Cellini.

In the world of art, this natural oxidation is a sign that the metal is corroding. Bronze statues and copper alloy artifacts must be protected, or the surfaces will corrode to some degree and can be visually affected as the surface metal combines to form the various corrosion salts. Sculpture, if not maintained, will oxidize and develop a patina. In severe cases, particularly in chloride environments, the copper alloy can undergo a destructive corrosion condition known as "bronze disease." However, one only has to look at the ancient sculptures that have been residing under the sea for centuries to observe that they may have corrosion products, but they are still intact and recognizable.

When some maintenance is performed, bronze sculptures can last centuries and appear as if they were cast in recent times. Figure 1.10 shows the famous sculpture *Perseus with the Head of Medusa*, created by Benvenuto Cellini around 1550 and located in the Loggia dei Lanzi in Florence, Italy. One of the most intricate and beautiful cast sculptures in existence, it stands at 5.2 m (approximately 18 ft.) as measured from the stone base to the top of Medusa's head. This remarkable, nearly 500-year-old sculpture is well maintained.

COPPER MINERALS

Copper and the brass and bronze alloys of copper are metals that have been with mankind since antiquity. Gold, silver, platinum, and copper are the only metals found in their native state. Copper, being less rare and occasionally found on the surface, was more available to early man. When the nature of the material was discovered and disseminated by our early ancestors, the Stone Age was over and the Copper Age was upon mankind. Useful tools that could be reused and reshaped originated with the metal copper.

Copper makes up only about 0.007% of the Earth's upper mineral strata. Like iron and nickel it is a denser metal, so one would expect copper to be found deeper in the Earth, as lighter elements such as silicon and aluminum would be expected to float over heavier elements (Table 1.2).

Primary copper ores are called "porphyry deposits." These ores are the virgin, nonrecycled sources of copper. They flow during magma releases but they often combine with sulfur to form heavy copper and iron sulfides, such as the minerals chalcopyrite ($CuFeS_2$), a mineral composed of copper, iron, and sulfur, and bornite (Cu_2FeS_4). Most copper is miles below the surface of the Earth. As the Earth was formed these heavier metals sank, leaving a surface with very little copper. These heavy compounds tended to sink deeper in the fluid flows and were then expelled in magma during an eruption.[3] This occurred in the Rocky Mountain region in the United States and in the formation of the Andes in South America.

Both of these regions are areas where large deposits of copper minerals are still being mined. Similar regions where volcanic activity over the eons moved minerals onto the surface of the planet exist around the world; places such as Papua province, where the Sudirman Mountains are located, and the southern Congo are rich in porphyry mineral deposits.

TABLE 1.2 Approximate percentages of elements in the Earth's upper mineral strata.

Element	Percentage (%)
Oxygen	46
Silicon	27
Aluminum	8.1
Iron	6.3
Nickel	0.009
Zinc	0.008
Copper	0.007

[3] *Geology* 66, no. 255 (2008); *Science* 319, no. 5871 (March 28, 2008); Rice University, "Copper Chains: Earth's Deep-Seated Hold on Copper Revealed," *ScienceDaily* (April 5, 2012).

TABLE 1.3 Mineral forms of copper.

Common mineral name	Formula	Color
Cuprite	Cu_2O	Red oxide
Tenorite	CuO	Black
Malachite	$CuCO_3(OH)_2$	Intense green with banding
Pseudomalachite	$Cu_5(PO_4)_2(OH)_4$	Emerald green
Azurite	$Cu_3(CO_3)_2(OH)_2$	Intense blue
Bornite	Cu_5FeS_4	Dark red with slight iridescence
Chalcocite	CuS	Black or black gray
Brochantite	$Cu_4SO_4(OH)_6$	Green
Antlerite	$Cu_3SO_4(OH)_6$	Green
Nantokite	$CuCl$	Pale green
Atacamite	$Cu_2(OH)_3Cl$	Crystalline green

The mineral form of a metal is more stable and slow to change. When exposed to the atmosphere copper and its alloys can develop an oxide layer that approaches one of the more common mineral forms of copper. Cuprite, for instance, forms on exposed copper alloy surfaces when subjected to heat and humidity. Nantokite with atacamite can form on copper alloys exposed to chlorides when near the sea (Table 1.3).

Bronze sculptures that have been exposed to the environment for a long period of time can form several of these mineral compounds, which can appear as spots or streaks on the surface of the original patina provided by the foundry (Figure 1.11).

HISTORY

The distinctive color copper possesses differentiates it from other natural minerals. The occasional pure native form of the metal would have attracted early man to this heavy, dense rock—a rock that would not fracture but that would yield to blows and take a different shape. Its weight and malleable nature made copper a useful discovery, and because of these attributes early man probably collected it when he came across it.

Easily recognizable by its color and tactile nature, the substance would have aroused the inquisitive nature of early humans. The surface of native copper allowed early man to shape it by hammering it with stone or wood. Initial hammering was probably performed using round stones or shaped logs, after which the substance was hammered into wooden forms. The plasticity of the metal would have been like no other material known. Forms of copper, cold hammered into jewelry, weapons,

FIGURE 1.11 Various minerals of copper.

and tools, have been found in Anatolia dating back to 9000 BCE. Skills used in making clay pottery, developed prior to the Copper Age, would have been adapted to this metal, although more force would have been needed and harder materials necessary to shape the copper into a useful form.

Once someone found that you could soften the metal further by heating it, these lumps of copper could be flattened to platelike forms, even blades. As hammering is repeated on the copper form it thins out as it is being shaped and the worked metal hardens, losing some of its elasticity. An edge can be sharpened on coarse rocks, making a blade form that can act as a tool or a weapon. A copper edge is short-lived, but the utility it would have offered and the ease of working the metal would have made copper an important early material.

Every early civilization used copper in art and decoration. From the Sumerians and Chaldeans in Mesopotamia, across Egypt, to present-day Turkey and on to India, copper was a part of early civilizations dating back nearly to 8000 BC. We know this by the character of the metal itself and its corrosion resistance. Artifacts made from copper exhibit only minor decay, even after all these centuries (Figure 1.12).

In the Upper Peninsula of Michigan on the shores of Lake Superior, in the area known as Keweenaw, ancient copper mines have been found along with copper tools.[4] Thought to date back as far as 4000 BC, these ancient mines show that people dug shallow mines in search of copper. Large copper boulders and mile-long veins were intermixed with general rock and rubble in this region.

[4]Charles Whittlesey, *Ancient Mining on the Shores of Lake Superior* (Washington, DC: Smithsonian Institution, 1863).

FIGURE 1.12 Approximate time periods of major copper activity in ancient times.

High-purity copper lumps were intermixed with the rubble on or near the surface. Surface copper, sometimes referred to as "drift copper," was deposited by the receding glaciers. These early inhabitants had no available means of cutting the large chunks of metal down, so they gathered the smaller, more manageable portions and left the large sections with some of their broken tools. Early Native Americans are believed to have ventured yearly to this site to gather copper to make ornamentation and other items. Old stone hammers and axes have been found where these early inhabitants attempted to carve off sections they could transport back to their villages. Some sections were simply too massive to transport. You can see one of the larger boulders of nearly pure copper, called the Ontonagon Boulder, in the Smithsonian's National Museum of Natural History. This large boulder weighs in at 1682 kg (3708 lb.). The Keweenaw Indians claimed it as a sacred object and they sought its return.[5] This boulder was one of the "proofs" given in 1843 for starting the mineral rush to the Upper Peninsula region of Michigan (Figure 1.13).

This region of the United States was invaluable in bringing America into the Industrial Revolution. It was the first site of a mad rush for mineral wealth and preceded the California Gold Rush by

[5]Office of Repatriation (2000). Executive Summary Assessment of a Request for the Repatriation of the Ontonagon Boulder by the Keweenaw Bay Indian Community.

FIGURE 1.13 Image of the Ontonagon copper boulder.
Source: Wikimedia Commons; public domain.

a number of years. From the early 1800s and for nearly the next 150 years, copper was heavily mined in this region. More than 6 billion kg are said to have been mined during this time. Artifacts have been recovered from these and other ancient civilizations still intact, demonstrating the diverse uses this metal was put to by early mankind.

The ductile nature of copper was one of the first characteristics of the metal that early man was able to use to his advantage. Copper was hammered thin and used in crude water-piping systems in early Egypt, the Romans used thin plates of copper to clad the roof of structures such as the Pantheon, and helmets and shields were created by artisans familiar with working with this pliable metal (Figure 1.14).

FIGURE 1.14 Roman helmets made from copper.

Early Egypt made extensive use of copper and copper alloys. Those in the ruling classes of Egypt used handmade mirrors of copper; small, thin razors of copper; and even colorful makeup made from mixtures derived from the copper ores of malachite and azurite. As far back as 3500 BC the Egyptians learned to mix alloying elements with copper to modify its color and improve specific properties. Copper mixed with tin melts at around 913 °C, while unalloyed copper requires a temperature of 1083 °C to melt. The Egyptians realized they could cast the molten metal into molds by adding specific amounts of tin. The lost-wax casting method was developed, and casting of intricate forms came into use along with purposeful alloying of metals.[6] For example, bronze is harder than copper; although this lack of ductility makes shaping bronze difficult, its hardness and durability allows it to hold an edge better than softer copper.

The Pyramid of Khufu in Giza (also known as the Great Pyramid of Giza), constructed around 2560 BC, is made from over 2,300,000 massive blocks of stone. The stones were cut using copper alloy tools that could be reformed and reused.

Some of the earliest copper mines were in Cyprus. The word "copper" originates from the Latin word *cuprum*, which itself is a contraction of a Latin term meaning "the metal of Cyprus" and which references the extensive mines found on the island. The metal became a significant source of trade in the region of the eastern Mediterranean.

On the Adriatic coast of Italy, down in the "bootheel," is the city of Brindisi, which was known for bronze trade and for bronze casting in ancient times. The Latin term *es Brundisium* ("from Brindisi") is thought to be the root of the word "bronze"—but the amount of trade in copper and copper alloys in this region attests to the importance of copper.

The Colossus of Rhodes, one of the Seven Wonders of the World described in classical antiquity, was made of bronze. This statue of the Greek god Helios was cast in bronze and erected over the entry to the port of Rhodes. It stood more than 32 m in height. Unfortunately, it collapsed during an earthquake in 226 BC. Nothing remains of it today other than the description.

Given the value of the metal and the technology to remelt cast objects and recast them, many bronzes of the Greek and Roman period were taken and repurposed. One can see the many pedestals where they once stood when visiting the ancient sites of Rome and Greece.

In Nepal the Pashupatinath Temple was built around the fifth century to honor the Lord Pashupatinath, an incarnation of the Hindu god Shiva. The temple, located on the riverbank in Kathmandu and sometimes referred to as the "copper temple," was decorated with copper ornamentation when it was built. Today it is clad in modern forms of copper.

Copper alloys have always played a part in cladding and adorning religious buildings in a variety of civilizations. In Japan, the Todaiji Temple is home to a massive cast Buddha. In the eighteenth century, over 400 tons of cast bronze was used to produce this 15-meter-tall Buddha (Figure 1.15).

[6] H. Coghlan, *Notes on the Prehistoric Metallurgy of Copper and Bronze in the Old World* (1951), 47–62.

FIGURE 1.15 Giant cast Buddha at the Todaiji Temple in Japan.

In many Christian churches the use of copper alloys is prevalent. Candleholders, baptismal fonts, and many other polished brass religious totemic items adorn churches and synagogues (Figure 1.16). Church roofs and steeples are commonly clad in thin skins of copper and have been since the technology for rolling thin sheets of the metal were introduced. The metal's resistance to environmental change and its perception as an elegant and special material make it ideal for surfacing roofs of churches.

In the late eighteenth century and up through the end of the nineteenth century, the center for Western production of copper and copper alloys was Great Britain. Mines in Cornwall and smelting operations in Swansea, Wales, have been a major source of copper since before the time of the Romans. Throughout the nineteenth century these mines were very active in gathering ore to be smelted and turned into sheathing for the wooden hulls of naval ships, and for their brass fittings and bronze cannons. Research and production facilities were set up in Birmingham, and here is where many early alloys of brass were invented.

In the sixteenth, seventeenth, and eighteenth centuries naval warfare resulted in much of the copper mined being commandeered for fashioning into sheathing for ships and for the production of cannons. Thick bronze—Gunmetal—castings were made for the massive warships of European armies, and for several centuries it was the largest single use of the metal. These castings were elaborately decorated on the outer surface, as shown in Figure 1.17.

FIGURE 1.16 Image of religious totemic items.

THE MODERN COPPER PRODUCTION PROCESS

At the beginning of the Copper Age, the copper collected was most likely surface mineral forms of the metal with few impurities. Called "native copper," this form has few impurities and requires no smelting. Early man would heat the metal to soften it, then hammer it into forms with wood or stones to arrive at useful shapes. The early Romans mined copper in the region known as Germania around 300 BCE. They heated the ore to drive out the sulfur that usually accompanies it. This would go on

FIGURE 1.17 Bronze cannon.

for several days, and eventually what is known as "matte copper" would be produced. There were several ores sought by ancient man that yielded the greatest amount of copper from this smelting operation. These ores were more readily identifiable by their color and appearance.

Primary ore	Color/structure
Cuprite	Reddish/oxide
Melaconite	Black/oxide
Malachite	Green/copper carbonate
Azurite	Blue/copper carbonite

Copper mining and copper production today involves strip mining where large ore deposits are found near the surface, such as the large strip mine in Arizona known as the Morenci mine and the Escondida mine in Chile. Deep deposits are mined by extending mine shafts to rich deposits, as in the case of the El Teniente mine in Chile.

The ore is extracted and crushed mechanically to produce smaller fragments, which are ground further and floated in a water slurry to concentrate the copper. This slurry is then heated or smelted to remove impurities such as sulfur. Nearly 99% of the sulfur is captured and converted to sulfuric acid, which is then used in other industrial processes. The acid is also used to leach copper from the ore, thus eliminating the need to smelt, saving energy and water. For every ton of copper produced, three tons of sulfuric acid is produced, as the sulfur dioxide is captured and converted to the acid for further industrial use.

These processes result in copper that is 98% pure. At this point the copper is called "blister copper." The blister copper is further refined by heating to remove oxygen and other contaminants. This copper is not suitable for electrical use, but it can be used for other processes. It is usually cast into large blocks called "anodes." These anodes are further refined to remove the traces of gold, silver, and other precious metals that often accompany copper mining. The anodes are sent to electrolytic cells where they are dissolved in an acid electrolyte and redeposited onto a cathode, leaving a copper with a 99% minimum purity. The copper is now called "electrolytic tough pitch" and is 99.94–99.96% pure. From here these cathodes of electrolytic tough pitch copper are remelted and cast into forms to be converted to plates, sheets, wire, rod, tubes, and extrusions (Figure 1.18).

New methods are being sought to reduce the environmental impact of copper processing. One such technique is electrowinning, which uses ores with low levels of copper—too low to make

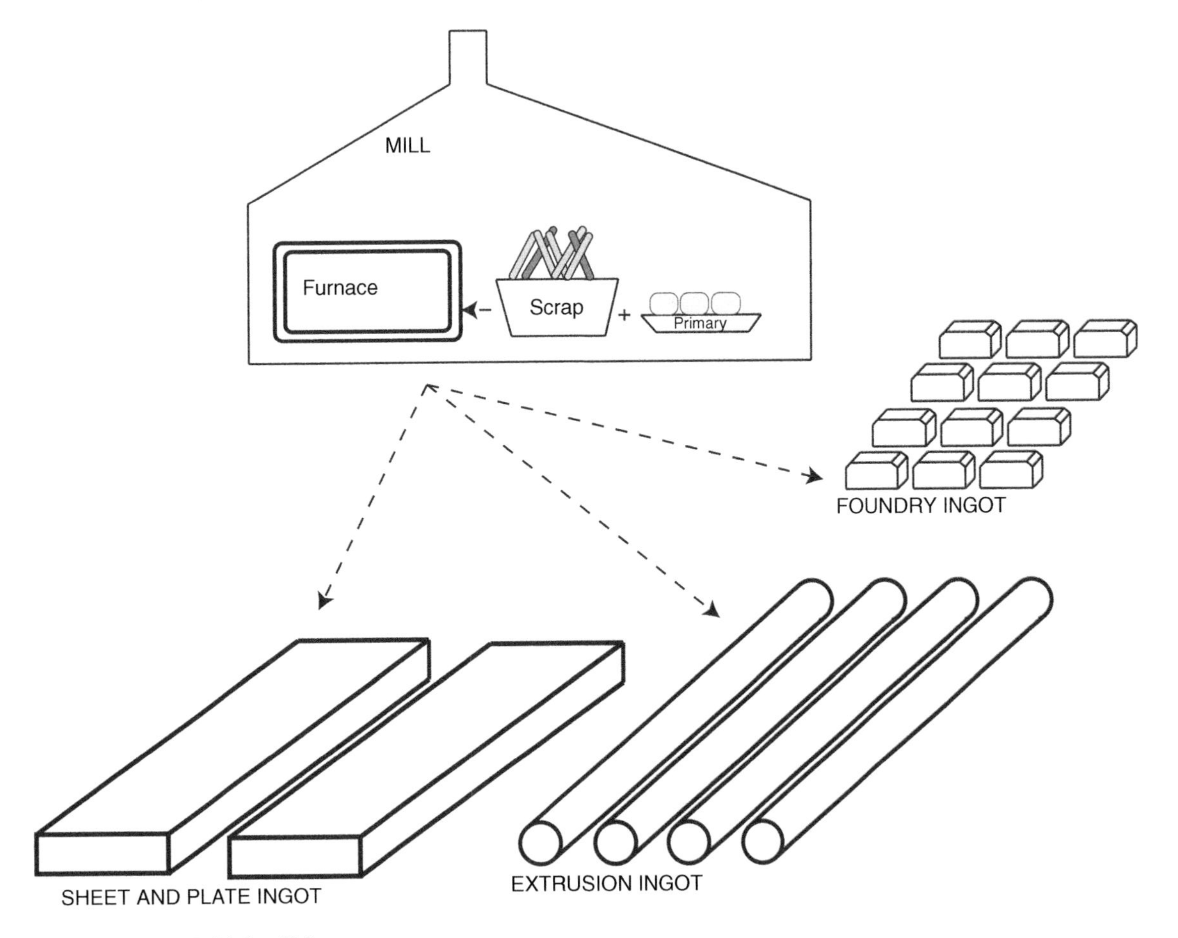

FIGURE 1.18 Initial mill forms.

smelting feasible—and crushes them into powder then dissolves them in an electrolyte. The copper is removed by electrolysis. Biomining, a technique that uses bacteria extraction to remove copper from low-grade sulfide ores, is another method that reduces the environmental impact of copper processing. Biomining (also known as bioleaching) is a much cleaner process than traditional leaching or smelting operations. In biomining, sulfide ores such as chalcopyrite (Cu_2S) undergo a multistep extraction process to remove sulfur and leave the copper ions. The microorganisms work with low concentrations of copper as they break down minerals into various elemental constituents. Mine tailings, once considered too difficult and costly to extract copper from, can be biomined.

Subsequent steps of copper production are similar to those of other metals.

SUSTAINABILITY, ENVIRONMENTAL, AND HYGIENIC CONCERNS

Sustainability involves striking a balance: a balance in which the activities involved in using a metal, such as copper, incorporate all aspects of the material's production and usage to arrive at a workable and sustainable accord with the future. This accord must consider the impact on the environment, the people working with the material, and the future generations that will be affected by the material. In some sense, copper can be a paradox. Undoubtedly mining has had an environmental impact, but copper also has had a significant impact in other areas where environmental impact is lessoned. The enlightenment of those in the industry with respect to our need to look beyond today is playing a major role in facing the question of sustainability.

There is no question that mining practices in the last century have negatively impacted the environment. Greed, mismanagement, and shortsighted ignorance have led mankind to scar the Earth in pursuit of higher profits. It is important to note, however, the change in thinking that is occurring in many of the largest mining firms.

One of the biggest environmental impacts of copper mining is the tailing pond used to contain the waste material that is the result of the extraction of ore. One of the largest and most toxic of these bodies of water can be found in Butte, Montana. In reality it is a sizable lake called the Berkeley Pit. This is a Superfund site that contains acidic waters and toxic chemicals left when the mining company closed the mine in the early 1980s. Today a pilot program is in place to remove, clean, and neutralize the water and to release it in a monitoring program. The expectation is that copper will be able to be extracted from the water by passing it through an iron filtering system.

This process of dewatering tailing ponds and flooded sites and recovering copper in the process is in place at a number of the largest mining operations around the world. The plan and the hope is that not only will this process capture a significant amount of copper, it will eliminate the dangerous water in these sites and return clean water back to the environment.

Copper is in higher demand today than ever before. The demand for this metal as technology advances in all areas, is growing. As we move away from the combustion engine and hydrocarbon energy sources, copper and copper alloys will play a greater role.

Something near 25 million tons of newly mined copper is produced each year. The rich deposits near the surface have long since been depleted. Today the ore is of lower grade, which means the yield is less. For every 100 kg of rock removed in a mining pit, about 1 kg of copper is obtained.[7]

[7]John Young, *Mining the Earth*, (Washington, DC: Worldwatch Institute, 1992).

As with other metals, copper production involves using a lot of energy, both for extraction and for refinement. However, today copper mining and production use less energy to extract copper due to process improvements.

COPPER AND WATER

There has been a lot of concern about runoff from copper surfaces entering streams and the human water supply. The fear is that these surfaces will allow the ionic form of copper (Cu_2^{+2}) to leach from the copper as water moves over the surface. There have been numerous tests to determine the level of toxicity generated from rainwater as it passes over the surface of a copper roof and is discharged into groundwater or the aboveground flow. Tests at the point of outlet do show toxicity, but as the copper interacts with the soil, water, and plant life it is combined and absorbed, eliminating the toxicity.[8] Copper ions and their toxicity are negated when exposed to the calcium and magnesium salts in hard water. Copper is absorbed rapidly into soils, in particular clay soil, as moisture containing copper ions passes through and over the surface.

Biological surveys taken of areas where water discharge has been occurring found no effects on living aquatic organisms from a prolonged exposure to copper runoff.[9] Laboratory tests, however, show copper concentrations of $14\,\mu g\,l^{-1}$ can have an adverse effect on the health of rainbow trout fry.[10]

It is important to understand that if you eliminate all means of copper ion absorption—soils, plants, and even hard water elements—a toxic situation may develop from the concentration of copper ions. However, in one natural setting where copper runoff collects in a rain garden pond, no adverse effects on amphibian, bird, reptile, insect, or plant growth have been observed for over two decades.[11]

An extensive study of the San Francisco Bay area was conducted between 1993 and 1999 to determine site-specific water quality criteria for copper and whether copper toxicity levels in estuary waters could be predicted using the biotic ligand model (BLM).[12] The tests analyzed the common blue mussel, *Mytilus edulis*, whose habitat is estuary waters. This species of mussel is considered to be the aquatic animal most sensitive to copper exposure in salt water by the US Environmental Protection Agency. Data was collected three times a year from 16 to 26 sites throughout the Bay

---

[8]Dale Peters, "New Research on Runoff from Copper Roofs," *Innovations*, Copper Development Association, April 1999.
[9]E. Grilli et al., *Unknown Nobleness* (KME and SMI, 2002), 24.
[10]Darlene Adrian et al., *The Hidden Environmental Impacts of Copper in the Christchurch Rebuild* (Christchurch, New Zealand: University of Canterbury, 2015), 18.
[11]The site is the author's pond. The author has conducted a yearly study of the aquatic life in the pond, which is immediately adjacent to runoff from copper roofs and copper walls that drain directly into the pond. Hundreds of frogs, intensive vegetation, dragonfly larvae, and hummingbirds have been observed each year.
[12]"A Screening Level Risk Assessment of Waterborne Copper in the San Francisco Bay Area, USA," in *Proceedings of the Copper 2003-Cobre 2003, The 5th International Conference-Volume II*, Santiago, Chile (November 30–December 3, 2003). Copper 2003: Health, Environment and Sustainable Development, VII, 531–541.

Area between1993 and 1999. The results of the findings from 487 site and data-specific observations concluded there was little risk in the estuaries tested.

As for potable water, consider that for decades water has been delivered to modern homes by means of copper tubing and brass fittings and forgings. There have been no known effects from copper tubing, and copper piping continues to be the material of choice for many homes. Ion release from copper tubing is greatest when first exposed to water flowing within it. After a short period of exposure, the air in the water combines with the surface of the copper tube, forming a layer of protection that reduces the release of copper.

Copper forms two oxides when exposed to water: cupric oxide (CuO) and cuprous oxide (Cu_2O). It is the cuprous oxide that acts as a biocide. Cuprous oxide is dark red in color. The size of the compound makes it a nonissue for humans. Cupric oxide is found in many over-the-counter vitamin supplements. It has been determined that the human body cannot absorb the copper from this compound.[13]

The World Health Organization (WHO) has stated that copper deficiency is more of a concern than copper toxicity for the general population as a whole.[14]

Humans, animals, and plants need trace amounts of copper. Copper is an essential nutrient, and the degree of copper intake by plant and animal systems is believed to be regulated by internal cellular activity. Plants need some trace amounts of copper to flourish. Copper is a fundamental component of the ability of a plant to properly function.[15]

For humans, copper acts as a defense against free radicals and is used in cellular respiration. Many foods we consume contain copper. Dark chocolate, for example, has more copper than your daily requirement.

HEALTH AND SAFETY

Exposure to copper dust and fumes created during processing and fabrication processes can irritate the mucous membranes of nasal cavities as well as the eyes and throat. Excessive and long-term exposure to dust and fumes from welding can cause dermatitis and pains in the lungs and chest.

Breathing fumes from copper alloy welding operations can cause upper respiratory problems, chills, aching muscles, nausea, and stomach and throat pain.

Copper itself is less toxic than the patina coating that develops on it. When working with copper and copper alloys, avoid breathing in copper salts, fumes from welding, and dust from grinding. Wear protective particle masks, gloves, and eye protection. General handling of copper does not pose a problem. Eye irritation can occur in hot and humid conditions when handling copper then touching the eyes or face.

[13] David H. Baker, "Cupric Oxide Should Not Be Used as a Copper Supplement for Either Animals or Humans," *Journal of Nutrition* 129, no. 12 (December 1999): 2278–2279, https://doi.org/10.1093/jn/129.12.2278.
[14] World Health Organization, *Guidelines for Drinking-Water Quality* (2004).
[15] E. Grilli et al., *Unknown Nobleness* (KME and SMI, 2002), 43–48.

COPPER: THE ANTIMICROBIAL METAL

In the context of health and hygiene, copper has an additional valuable characteristic: it has long been suggested that copper has the ability to aid in the combatting of human ailments. For example, some people wear copper bracelets to combat arthritis—although there is substantial doubt about whether such jewelry really helps.

The Aztec Indians supposedly gargled with a solution that contained copper to reduce the effects of a sore throat. And in the first century AD, Pliny the Elder used a treatment of copper oxide mixed with honey to eradicate intestinal worms.

In the last several decades there has been substantial testing of the efficacy of various surfaces and wipes in eliminating harmful bacteria, and copper and copper alloys have performed extremely well.

A study in 1983 on microbial growth on doorknobs[16] found that hospital door handles made of copper alloys had no growth in microbes on the surface, while other door handles made of stainless steel showed significant microbial growth. Since then there have been a number of studies on the efficacy of copper and copper alloys in combatting microbial growth.

The University of Southampton in the United Kingdom has performed studies where the growth of microbes on various surfaces was measured. The studies found that copper surfaces killed staphylococcus and enterococci bacteria in 2 hours or less. In fact, there was a significant drop in cultures within the first 45 minutes.[17]

The tests were performed at ambient temperatures. The US Environmental Protection Agency reviewed the study and concluded that copper is a material that can benefit public health by killing harmful bacteria. Further tests have been performed on *E. coli* and *Legionella* bacteria and results were similar.

Since coating the copper in any way can reduce the antimicrobial benefit, this is where the main constraint lies. Copper alloys will tarnish from humidity and moisture on the surface and from fingerprints. As a result, polished surfaces seem to perform the best in conveying antimicrobial benefits because they offer little foothold for bacteria and are easier to clean. In contrast, satin finishes hold oils and moisture and will spot from tarnishing.

There are several alloys designed to be beautiful that also offer antimicrobial benefits. These are more expensive due to their custom nature; however, they offer beauty and a reduced affinity for tarnishing and oxidizing. They are easier to clean. You can remove most fingerprints with isopropyl alcohol (Figure 1.19).

These beautiful alloys of copper contain nickel, iron, zinc, and aluminum in varying levels. They have been tested with the goal of fabricating a metal that does not promote bacterial growth and is proven to kill viruses and bacteria.

---

[16] P. J. Kuhn, "Doorknobs: A Source of Nosocomial Infection?," *Diagnostic Medicine* (1983).

[17] J. O. Noyce et al., "Potential Use of Copper Surfaces to Reduce Survival of Epidemic Methicillin-Resistant *Staphylococcus aureus* in the Healthcare Environment," *Journal of Hospital Infection* 63 (2006): 289–287.

FIGURE 1.19 Examples of special-order copper alloys with biocide properties from KME.

To convey their antimicrobial benefits, copper alloys must not be coated in wax or lacquers. There also needs to be a maintenance procedure to eliminate fingerprints and other grease or oils that will mar their inherent beauty.

COPPER ALLOYS FOR THE ARTS

Bronze Sculpture

Bronze has been a stalwart of artists and art collections for centuries.

The most common form is cast bronze statuary; although there are other forms of bronze used for art, such as wrought forms of sheet or plate, casting is the process used for the majority of work.

Techniques for casting bronze have been perfected by foundries over the years, with by far the most common being lost-wax casting. There have been several material advancements in the lost-wax casting method, but the process has been virtually the same for the past 5000 years.[18]

---

[18]L. B. Hunt, "The Lost History of Lost Wax Casting," *Gold Bulletin* 13, no. 2 (1980): 63–79, https://doi.org/10.1007/BF03215456.

In the lost-wax method, a sculpture is created in clay and a mold is taken from this clay model to create a shell. The inside of the shell is an accurate negative representation of the clay model. The shell is coated with a layer of wax whose thickness is the same as the final casting. In ancient times beeswax from wild bees was used.

Into the wax shell is placed refractory material to form the core of the casting. The sprues and vents are made of wax tubes that are connected to the wax mold, and it is through these tubes that the molten metal will enter the mold. The entire assembly is coated with a refractory slurry that will harden around the wax, wax vents, and tubes.

The assembly is heated and the wax melts out, leaving a space to be filled with the molten metal. The molten metal is poured into the mold and fills the cavity where the wax once was. Once cooled the metal is removed from the mold and finished to remove refractory metal, to fill porosity, and to remove the feeder tubes and venting. Usually a patina is applied to the finish work. Figure 1.20 shows an intricate cast dragon adorning the Town Hall in Munich.

FIGURE 1.20 Cast dragon ornamentation on the Town Hall in Munich, Germany.

Beyond Casting

Bronze sculpture is still one of the most common uses of copper in art, but the metal lends itself to many other forms. Sheets material can be hammered and shaped and is a favorite metal for the chasing and repose work. Copper in particular is well suited for hammering into forms. Copper's ductility and its ability to elongate without cracking and to be softened by annealing make it well suited for shaping. Copper work hardens slowly, allowing for significant stretching and shaping as it undergoes plastic deformation without breaking (Figure 1.21).

This ductile characteristic of copper and many of its alloys allows for beautiful forms, textures, and shapes in artistic expression. Working with sheet copper alloys, artists can create significant and beautiful surfaces by patinating the surface. The metal reacts with other compounds and can form beautiful decorative patinas that offer protection from further atmospheric exposure, or it can be prepatinated to develop its natural beauty from the beginning. Figure 1.22 is a large hammered wall of copper that has been patinated and protected with a clear coating.

Many of the techniques used in art can be expanded into architectural surfacing and expression. This incredible metal offers the design community a material for creating timeless beauty.

FIGURE 1.21 A repoussé copper piece from Jerusalem.

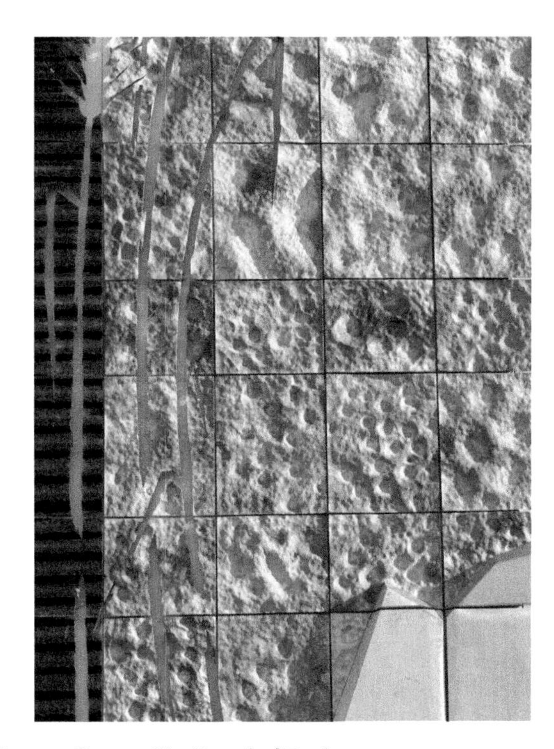

FIGURE 1.22 *Heartland Harvest*, Kansas City Board of Trade.

Copper and copper alloys, when properly installed and maintained, will age gracefully in exterior applications—or they can be frozen with clear protective coatings of lacquers specifically created to protect copper.

Copper Relative to Other Architectural Metals

Compared to other metals available for art and architecture, copper is considered one of the relatively more expensive materials. Like gold and silver it is traded as a commodity on the marketplace and thus it is exposed to speculation and the forces of supply and demand. The price of copper fluctuates as these forces influence its price. Figure 1.23 shows the fluctuations of copper spot prices in US dollars per pound over a 30-year time period. The price of the copper alloys correlates to the price of copper and is adjusted by the alloying element price and also by the cost of and demand for producing the alloy.

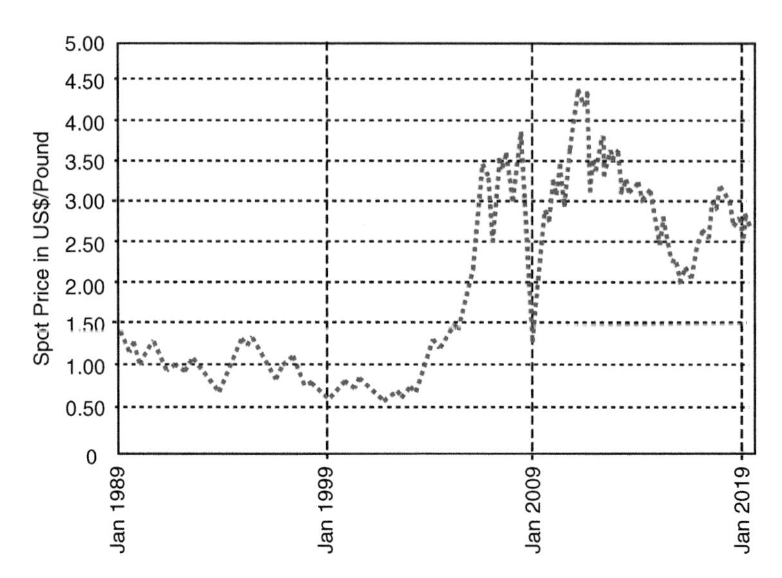

FIGURE 1.23 Copper spot price in US dollars per pound over a 30-year period.

The price fluctuation is not unlike other metals that are traded on the commodity future markets. Copper is an industrial metal and the effects of speculation and hedging is felt in the price movement.

When comparing the various attributes of the metals used in art and architecture, copper has its advantages and disadvantages depending on the designer's taste. The wide range of attributes this metal possesses makes it worth considering in many applications. Appendix A has comparisons of various attributes of metals.

CHAPTER 2

Copper and Its Alloys

INTRODUCTION

We often think of copper, brass, and bronze as completely different metals. In many ways they are. In the context of art and architecture, appearance and color are usually foremost, followed in no particular order by corrosion resistance, formability, and cost. There is a vast array of alloys in manufacture with select properties created for a specific environment or manufacturing process. That is the beauty of a metal that has been a mainstay to mankind for centuries. All sorts of endeavors have been devised around this highly versatile metal.

We think of copper as a soft, malleable metal, and in its commercially pure, annealed state it is. However, when alloyed with certain other elements and strengthened by thermal and mechanical means some alloys approach the strength of steels. Copper alloyed with beryllium has a tensile strength of nearly 1320 MPa (191,500 psi), with a hardness of 370 HB (Brinell hardness). Between the soft, ductile annealed copper and the beryllium alloy, there are a vast number of alloys for use.

In art and architecture there are a select number of copper alloys that are in common use and several not so common. As with other metals there are constraints in the production of copper alloys at the Mill source. Some of the more common alloys are produced on a regular basis, while the more specific alloys require a sizable production run, making them less available for specific use on smaller architectural or art projects. The recommendation is to check with reputable and knowledgeable suppliers to confirm availability for a particular design.

THE RICH HISTORY OF COPPER ALLOYS

The term "copper" is used to refer to copper alloys that are at least 99.3% pure copper. In contrast, the term "high-copper alloy" refers to a member of another group of alloys that contains no less than 94% copper, while "copper alloy" refers to alloys composed of 50–94% copper. These major groupings apply to both wrought and cast forms of copper.

Copper alloy production and use has such a rich history. Over the centuries a vernacular all its own has surrounded the metal. Borne from industries that worked with copper and copper alloys, such names as Admiralty Metal, Engravers Brass, Naval Brass, Jewelry Bronze, Gilding Metal and Cartridge Brass came into use to describe a particular alloy established for an industry. (Table 2.1).

TABLE 2.1 A few of the names given to alloys of copper.

	Approximate percentage: content varies			
Name	**Cu (%)**	**Zn (%)**	**Other**	**Closest alloy**
Abyssinian Gold	90	10		C22000
Aich's Metal	60	38.2	1.8% Fe	C37000
Admiralty Brass	71	28	1% Sn	C44300
Aluminum Brass	77.5	20.5	2% Al	C68700
Architectural Bronze	57	40	3% Pb	C38500
Bell Metal	78	22		C24000
Beryllium copper	98.3		1.7% Be	C17200
Cartridge Brass	70	30		C26000
Delta Metal	61.2	27.4	0.18% Fe, 0.36% Pb	C46700
Dutch Gold	82	18		C24000
Engravers Brass	61.5	34.5	2% Pb	C35300
Everdur	94	0.25	4% Sn, 1.5% Mn	C87300
German Silver	60	20	20% Ni	C75200
Gilding Metal	95	5		C21000
Gunmetal	88	4	8% Sn	C90500
Jewelry Bronze	87.5	12.5		C22600
Manganese brass	70	28.7	1.3% Mn	C66700
Muntz Metal	60	40		C28000
Naval Brass	60	39.2	0.8% Sn	C46400
Nickel silver	65	25	10% Ni	C74500
Nordic Gold	89	5	1% Sn	C22000
Prince's/Prince Rupert's Metal	75.7	24.3		C24000
Phosphor bronze	92		8% Sn	C52400
Tobin Bronze	61.2	27.4	0.18% Fe, 0.36% Pb	C46700
Tonval Brass	58.5	35.9	0.30% Fe, 1.5% Pb	CW617N
Silicon tombac	80	16	4% Si	CC761S
Silicon bronze	97		3% Si	C65500
Zirconium copper	99.8		0.20% Zr	C15000

The metal and the alloying constituents became engrained. Colorful names such as Muntz Metal, German Silver, Dutch Gold, and Prince's (or Prince Rupert's) Metal, are still in the lexicon of the ornamental metal worker. It doesn't stop there. As modern production and specialized alloys came into use, descriptive names such as beryllium copper, zirconium copper, phosphor bronze, silicon bronze, aluminum brass, and manganese brass became commonplace.

These names were borne from the industries that worked with the metal. Throughout the nineteenth century and into the mid-1900s companies specialized in working solely with particular copper alloys. The craftspeople had their own jargon for the custom metal alloys, and often the name referred to the end use or form. Names such as Gunmetal, Bell Metal, and Jewelry Bronze referred to the end products these alloys were used in. The makeup of an alloy often had a liberal range of allowable variance in constituents but was still sufficiently distinct from other copper alloys.

Admiralty Bronze, or Admiralty Metal, was developed by the British Admiralty around 1876. In the early 1800s, Swansea, Wales, became the site of the largest mining operations in the world and Birmingham, England, was one of the largest producers of wrought copper alloys in the world at that time. The British navy found that using copper alloy as sheathing for the undersides of its ships kept the *Teredo* shipworm from burrowing into the wooden hulls and also deterred the growth of barnacles, which normally had to be scraped from the hulls.

Similarly, Aich's Metal, named after its inventor, had a small amount of iron alloyed with copper and zinc. When heated it could be forged without cracking. Like Admiralty Brass, it was developed for use by the British navy in 1860.

ELEMENTS ADDED TO COPPER

There are a number of elements added to copper that provide specific characteristics for industry. For art and architecture, the following elements can be found in many of the alloys used. The quantity of the alloying element is important for metallurgic reasons as well as mechanical and esthetic reasons.

Symbol	Element	Effect
Zn	Zinc	The most common alloying element added to copper. As zinc is added, the strength and resilience increases. As zinc is added, the color becomes more yellow and golden.
Si	Silicon	Reduces the melting point while increasing fluidity in castings. Improves machinability. Deoxidizes copper but slight reduction in corrosion resistance.
Sn	Tin	Improves corrosion resistance and hardness. The tin – copper bronzes are some of the most corrosion resistant of the alloys.
Al	Aluminum	Improves strength and corrosion resistance. Increases hardness. Color is a yellow – gold tone. Improves fatigue resistance.
Mn	Manganese	Improves casting properties and enhances the strength.
Pb	Lead	Added to aid in machining copper alloys. Lead is insoluble in copper and forms clumps around grain boundaries. Machining is improved dramatically.

P	Phosphorus	Added in small quantities to deoxidize copper, improves ductility. Increases strength and corrosion resistance. Improves resilience.
Ni	Nickel	Improves hardness and strength. Enhances corrosion resistance significantly.
Ag	Silver	Not added as an alloying element but traces of silver can be in some copper and copper alloys. Silver improves electrical conductivity.
As	Arsenic	Not added to alloys but is often found with copper in some ores. Arsenic bronze and arsenic copper has been used in sculpture from antiquity.
Fe	Iron	Adding small amounts of iron to the copper-nickel alloys improves corrosion.

ALLOY DESIGNATION SYSTEM

Each form of copper alloy, both wrought and cast, has a specific designation. Because of the various names associated with an alloy's particular use or particular industry, and because the associated alloying constituents varied from one producer to another, a more refined numbering system was established to provide consistency. In this book we will use the alloy designation established by the Unified Numbering System (UNS) for Metals and Alloys.[1] It has a logical format and industry has weighed in on the allowable tolerances. Constituents that are given are nominal amounts and should be verified with the Mill producer.

Appendix C discusses how other regions of the world use alternative designations of the copper alloys. Some of these are cross-referenced in the alloy descriptions used in this chapter.

THE UNIFIED NUMBERING SYSTEM

Under the UNS, the prefix "C" followed by a series of digits that designates the family of similar alloys with differing levels of similar alloying elements is used for copper alloy designations.

The wrought alloy designations begin with a numeral between 1 and 7 depending on the alloying components, while cast alloy designations begin with an 8 or a 9. These numbers are similar to numbers used previously by the Copper Development Association (Tables 2.2 and 2.3).

Of all these various alloys, there is but a few used regularly in art and architecture. The Mill source, the stocking house, the form of the metal, and the foundry confer production constraints that dictate the alloy on hand. The following sections provide basic descriptions of those alloys and a few that could offer interesting alternatives for use in art and architecture.

[1]The Unified Numbering System for Metals and Alloys is jointly managed by the American Society of Testing and Materials (ASTM) and the Society of Automotive Engineers (SAE).

TABLE 2.2 UNS number ranges for wrought copper alloys.

Wrought alloys	UNS number range	Main alloy elements
Copper	C10100–C15760	No less than 99% copper
High-copper alloys	C16200–C19600	No less than 96% copper
Brass	C20500–C28580	Zinc
Leaded brass	C31200–C38590	Zinc, lead
Tin brass	C40400–C49080	Zinc, tin, lead
Phosphor bronze	C50100–C52400	Tin, phosphorus
Leaded phosphor bronze	C53200–C54800	Tin, phosphorus, lead
Copper–silver phosphorus	C55180–C55285	Phosphorus, silver
Aluminum bronze	C60600–C64400	Aluminum, nickel, iron, silicon, tin
Silicon bronze	C64700–C66100	Silicon, tin
Other copper and zinc	C66400–C69900	—
Copper–nickel alloys	C70000–C79900	Nickel, iron
Nickel–silver alloys	C73200–C79900	Nickel, zinc

TABLE 2.3 UNS number ranges for cast copper alloys.

Cast alloys	UNS number range	Main alloying elements
Copper	C80100–C81100	Greater than 99% copper
High-copper alloys	C81300–C82800	Greater than 94% copper
Red and leaded red alloys	C83300–C84800	Zinc, tin, lead; 75–89% copper
Yellow and leaded yellow	C85200–C85800	Zinc, tin, lead; 57–74% copper
Manganese bronze	C86100–C86800	Zinc, manganese, iron, lead
Silicon brass and bronze	C87300–C87900	Zinc, silicon
Tin bronze	C90200–C94500	Tin, zinc, lead
Nickel–tin bronze	C94700–C94900	Nickel, tin, zinc, lead
Aluminum bronze	C95200–C95810	Aluminum, iron, nickel
Copper–nickel alloys	C96200–C96800	Nickel, iron
Nickel silvers	C97300–C97800	Nickel, zinc, lead, tin
Leaded copper	C98200–C98800	Lead
Special alloys	C99000–C99999	White manganese brass and others

TEMPERS

Although there are a number of tempers possible for copper alloys, only a few find their use directly in art and architecture. Tempers for the wrought forms are produced in the general categories of hot rolled, cold rolled, and annealed tempers. Other thermal-mechanical methods are used on some of the alloys to increase their strength when the forms of the copper alloys are other than the wrought forms of sheet and plate.

The names for tempers come from the same world that created the names for alloys. Descriptors such as quarter-hard and half-hard are used to describe developed mechanical properties in wrought products. For example, the mechanical properties achieved in a quarter-hard temper of alloy C22000 are different from those achieved in a quarter-hard temper of alloy C11000 or alloy C26000. The ambiguous nature of temper names has been addressed by the Copper Development Association (CDA) by assigning an alphanumerical value to each temper (Tables 2.4, 2.5, and 2.6).

TABLE 2.4 A few common cold rolled wrought temper designations.

CDA designation	Description
H00	1/8 hard
H01	1/4 hard
H02	1/2 hard
H03	3/4 hard
H04	Full hard
H06	Extra hard
H08	Spring
H10	Extra spring
H12	Special spring
H13	Ultra spring
H14	Super spring
H50	Extruded and drawn
H52	Stamped and drawn
HR01	1/4 hard and stress relieved
HR02	1/2 hard and stress relieved
HR04	Full hard and stress relieved
HR08	Spring hard and stress relieved
HR10	Extra spring and stress relieved

Source: Active Standard ASTM B601.

TABLE 2.5 A few common hot rolled temper designations.

CDA designation	Description
M20	As hot rolled
M30	As hot extruded

Source: Active Standard ASTM B601.

TABLE 2.6 A few common temper designations of copper.

CDA designation	Description	Tensile strength		Rockwell HRF
		MPa	ksi	
O60	Soft temper	221	32	40
H00	1/8 hard	248	36	60
H01	1/4 hard	262	38	70
H02	1/2 hard	290	42	84
H04	Hard	310	45	90

Abbreviation ksi = kilopound per square inch.

Tempering of copper alloys involves cold working and heating. In wrought forms (sheet and plate in particular) the metal is passed through a series of rolls (called a "tempering mill") while either hot or cold. In cold rolling, the metal ribbon is rapidly passed between rolls that elongate the metal to a predetermined level. Each pass incrementally stretches the metal and the grains become smaller. This improves the surface quality and increases the yield strength of the metal. This also hardens the surface—thus the industry jargon eighth-hard, quarter-hard, and so on. The harder the surface, the more strength is imparted, which can have an effect on the downstream feasibility of shaping the metal.

Annealing is used on certain alloys to reduce or remove the hardness in a wrought form. Annealing involves the controlled heating of the metal to a point where the strained metal grains induced into the metal from cold working are disintegrated and re-form into other grains with the internal strain released. The heating causes the grains to move to regions of lower strain. As the temperature increases the grain size increases, and ductility is restored into the metal.

Another temper designation is the "as fabricated" temper, which refers to the temper achieved during the fabrication process. Most cast copper alloy forms used in art and architecture are provided in an as fabricated temper.

WROUGHT COPPER ALLOYS

Commercially Pure Copper

As mentioned previously, the name "copper" refers to copper alloys that are at least 99.3% pure. The most common wrought copper alloy used in art and architecture is the commercially pure alloy C11000, which is also one of the alloys known as "electrolytic tough pitch (ETP) copper." ETP copper refers to any copper that has been refined by electrolytic deposition. The process was developed in 1864 by a chemist in Birmingham, England.

When copper is remelted and cast it is electrolytically refined to reduce impurities to less than 0.1%. This removes all but trace impurities and a small amount of oxygen and yields the form Cu_2O. More than half of the copper produced is the ETP alloy. The most common trace impurities remaining in the ETP copper alloy are antimony, arsenic, bismuth, iron, lead, nitrogen, oxygen, silicon, selenium, and tellurium.

Oxygen affects copper mainly by impairing the ductility of the metal, so phosphorus or silicon is often added as a deoxidizer of the copper.

COPPER ALLOY C11000

UNS C11000	ASTM B152 ASTM B370 ASTM B248 ASTM B601
110	

The balance consists of approximately 0.04% oxygen and trace amounts of other elements.

Copper alloy C11000 is the most common of the ETP copper alloys. Its modulus of elasticity is 17 ksi (117 Mpa) and it is the alloy used in general architectural work. Copper roofing, copper cladding, and copper handrails are all created from this alloy (Figures 2.1 and 2.2). This ductile form is easily cold worked and can be very malleable when heated to a red intensity where it will become soft and pliable. All wrought forms are available in this alloy.

Available Forms

Sheet
Plate
Rod
Tube
Pipe
Rod
Extrusion

FIGURE 2.1 Copper handrail and copper wire rope.

FIGURE 2.2 Panel system made of the C11000 alloy at the de Young museum in San Francisco.

There are other copper alloys similar to the C11000 alloy that have additional alloying elements to create particular properties of value to specific industries. These have been created mostly for the electronics industry, where conductivity is critical, but some are alloys created for air conditioning and piping that aid in making sound brazed or welded connections. A few of these alloys are listed in the following table.

Alloy C12200 is another common alloy that is deoxidized. This alloy is considered for use where welding and brazing operations are required. Often called DHP (deoxidized high phosphorus) copper, the alloy comes in tubes and sheet and is similar to C11000.

UNS designation	Description	Composition	Application
C10100	Oxygen free	99.99% copper	Electronics
C10200	Oxygen free	99.95% copper	Electronics
C10300		99.95% copper with traces of phosphorus	
C10400	Oxygen free with silver	99.95% copper with traces of silver	
C10500	Oxygen free with silver	99.95% copper with traces of silver	
C10700	Oxygen free with silver	99.95% copper with traces of silver	
C10800		99.95% copper with phosphorus	
C12200	Phosphorus copper	99.90% copper	Gas and refrigerant lines
C14500	Tellurium copper	99.50% copper with tellurium	Machining
C18200	Chromium copper	99.10% copper, 0.9% chromium	Welding
C18700	Leaded copper	99.00% copper, 1.0% lead	Electrical

The mechanical properties of these alloys are similar, but they are specifically designed for a particular industry. Phosphorus is added to remove oxygen and to arrive at what are known as deoxidized phosphor coppers. Approximately 40% of the copper made is of this form. These are not used in electronics but in general copper work on heat exchange equipment.

Oxygen free copper is produced in controlled ovens with special atmospheres that keep the oxygen out of the initial cast block or billet of metal. This form of copper has trace amounts of oxygen and finds use in electronics. These pure alloys have very low electrical resistance and are used primarily to create electronic components. The standard used to gauge a metal's conductivity is called the International Annealed Copper Standard, or IACS. Pure oxygen free copper has a 100% IACS. Aluminum, by comparison, has a 61% IACS, while silver has a 105% IACS.

Approximately one third of the copper that is produced goes into producing alloys for use in the numerous industries that need specific properties of ductility, corrosion resistance, and esthetics. The following sections describe the copper alloys that find their way into art and architecture.

BRASSES

Brass is the name given to copper alloys with zinc as the major alloying constituent.

Over the years the terms "brass" and "bronze" have come to have less to do with the alloying constituents and more to do with the particular jargon developed around their use.

Brasses and bronzes make up a considerable amount of the copper alloys used in art and architecture. As compared to commercially pure copper, the zinc increases the strength and stiffness of the copper alloy without having a major effect on corrosion resistance. Brasses are valued for their unique, pleasing colors—colors that range from a golden bronze to a yellowish gold hue. If not protected, the pleasing surface hues can develop oxides on the surface that detract from the original beauty in a matter of days after exposure to the elements.

Often these brass alloys are darkened or highlighted to produce semitransparent, rich copper oxides and sulfides on the surface, then sealed with a clear protective coating.

There are a number of brass alloys. The following are the more common alloys used in art and architecture. Each is described in detail. The brasses listed are divided into groups, beginning with what are considered the true brasses and followed by the leaded brasses (Table 2.7).

TABLE 2.7 True brasses.

UNS designation	Zinc (%)	Name	Color
C21000	5	Gilding Metal	Coppery red
C22000	10	Commercial Bronze	Golden bronze
C23000	15	Red brass	Reddish gold
C24000	20	Low brass	Reddish gold
C26000	30	Cartridge Brass	Yellowish gold
C27000	35	Yellow brass	Golden yellow
C28000	40	Muntz Metal	Yellowish gold

COPPER ALLOY C21000

UNS C21000

210
Gilding Metal

ASTM B36
ASTM B248
ASTM B134
ASTM B587
ASTM B601

Alloying constituents added to Copper

5%

UNS designation	Temper	Tensile strength	Yield strength	Elongation % (2 in.–50 mm)	Rockwell Hardness B
C21000	H01	42 ksi (290 MPa)	32 ksi (221 MPa)	25	38
C21000	H02	48 ksi (331 MPa)	40 ksi (276 MPa)	12	52

Available Forms

Strip
Tube
Wire

Alloy C21000, also known as Gilding Metal, is not commonly used in art and architecture and is not readily available. The most common form is narrow strip. Its color is not much different from that of copper. The added zinc makes it harder and stronger than copper but not to a significant degree.

COPPER ALLOY C22000

UNS C22000

220
Commercial Bronze

ASTM B36
ASTM B248
ASTM B134
ASTM B587
ASTM B601

Alloying constituents added to Copper

10%

UNS designation	Temper	Tensile strength	Yield strength	Elongation % (2 in.–50 mm)	Rockwell Hardness B
C22000	H01	45 ksi (310 MPa)	35 ksi (241 MPa)	25	42
C22000	H02	52 ksi (358 MPa)	45 ksi (310 MPa)	11	58

Available Forms

Sheet
Plate
Rod
Tube
Wire

Copper alloy C22000, commonly known as Commercial Bronze, is a popular architectural metal. Coveted for its soft, golden bronze tone, C22000 can be mirror polished, glass bead blasted, or satin finished. It can also be highlighted chemically to create enhanced statuary surfaces. This copper alloy has excellent corrosion resistance, good strength, and ductility. C22000 is not available in shapes such as angles or channels, and bar is cut from thick plates if needed. Used in both interior and exterior applications, C22000 is a one of the more attractive wrought surfaces used in art and architecture (Figure 2.3).

COPPER ALLOY C23000

UNS C23000

230
Red Brass

ASTM B36
ASTM B248
ASTM B134
ASTM B587
ASTM B601

Alloying constituents added to Copper

15%

UNS designation	Temper	Tensile strength	Yield strength	Elongation % (2 in.–50 mm)	Rockwell Hardness B
C23000	H01	50 ksi (345 MPa)	39 ksi (269 MPa)	25	55
C23000	H02	57 ksi (393 MPa)	49 ksi (338 MPa)	12	65

Available Forms

Sheet
Rod
Tube
Wire

FIGURE 2.3 Copper alloy C22000 in the Ohio Holocaust and Liberators Memorial, Source: designed by Daniel Libeskind.

Copper alloy C23000, commonly referred to as red brass, is a common architectural alloy. The color has more yellow or golden tones than Commercial Bronze (alloy C22000). Alloy C23000 has good strength and is common in small stampings and formed trim. It is not available in shapes such as angles, extrusions, channels, or bars. Similarly to C22000, it can be polished, glass bead blasted, satin finished, and chemically oxidized to enhance the appearance.

COPPER ALLOY C24000

UNS C24000

240
Low Brass

ASTM B36
ASTM B134

Alloying constituents added to Copper

20%

UNS designation	Temper	Tensile strength	Yield strength	Elongation % (2 in.–50 mm)	Rockwell Hardness B
C24000	H01	53 ksi (360 MPa)	40 ksi (275 MPa)	30	55
C24000	H02	61 ksi (421 MPa)	50 ksi (345 MPa)	18	70

Available Forms

Strip

Sheet

Wire

Copper alloy C24000, also known as low brass, is in common use. It has good strength and decent corrosion resistance. It has a tint that tends toward golden yellow. Once used more frequently for architectural surfaces, today it is mostly the metal of musical horn instruments. This alloy has very good ductility and will take a lot of deep drawing and stretching.

COPPER ALLOY C26000

UNS C26000

260
Cartridge Brass

ASTM B36
ASTM B134
ASTM B587
ASTM 129

Alloying constituents added to Copper

30%

UNS designation	Temper	Tensile strength	Yield strength	Elongation % (2 in.–50 mm)	Rockwell Hardness B
C26000	H01	54 ksi (372 MPa)	40 ksi (276 MPa)	43	55
C26000	H02	62 ksi (427 MPa)	52 ksi (358 MPa)	25	70

Available Forms

Strip
Sheet
Plate
Bar
Tube
Wire

Alloy C26000 is commonly known as Cartridge Brass. This is the copper alloy used to make ammunition components, such as cartridges for bullets: thus the name. It has good strength, approaching that of some mild steels. Architectural uses are grilles and panels where the golden yellow, brass color is desired. It will take a mirror polish well to bring out the deep golden yellow (Figure 2.4).

FIGURE 2.4 Mirror polished C26000 alloy.

COPPER ALLOY C27000

UNS C27000

270
Yellow Brass

ASTM B36
ASTM B134
ASTM B587

Alloying constituents added to Copper

35%

UNS designation	Temper	Tensile strength	Yield strength	Elongation % (2 in.–50 mm)	Rockwell Hardness B
C27000	H01	54 ksi (372 MPa)	40 ksi (275 MPa)	43	55
C27000	H02	61 ksi (421 MPa)	50 ksi (345 MPa)	23	70

Available Forms

Sheet

Plate

Rod

Tube

Alloy C27000, commonly known as yellow brass, has a distinctive yellow color. This alloy is readily available in tube and rod forms, but less so in plate and sheet. It is a decorative alloy with a strength similar to C26000. Grilles and etched plates are common forms of this alloy in architecture.

Alloy C27200 is a similar form used to produce drawn tubes. Trace amounts of lead may be added to facilitate the drawing of the tube form.

Alloy C27450 is also a common yellow brass alloy. It too has lead and iron in amounts no greater than 0.25% and 0.35%, respectively. It is used as a rod and bar material.

COPPER ALLOY C28000

UNS C28000

280
Muntz Metal

ASTM B36
ASTM B135
ASTM B601

Alloying constituents added to Copper

40%

UNS designation	Temper	Tensile strength	Yield strength	Elongation % (2 in.–50 mm)	Rockwell Hardness B
C28000	H01	60 ksi (413 MPa)	35 ksi (241 MPa)	30	55
C28000	H02	70 ksi (482 MPa)	50 ksi (345 MPa)	10	75

Available Forms

Strip
Sheet
Plate
Rod

Alloy C28000 is a common architectural alloy known as Muntz Metal. This alloy is significantly more difficult to cold work. Cold forming this alloy can be challenging due to its strength and hardness. The edges of tight forms made from C28000 can crack at folds when cold formed. If significant shaping is required, then hot forming C28000 is suggested. Muntz Metal is an attractive golden alloy

FIGURE 2.5 Polished C28000 alloy entryway.

that can be mirror finished, satin finished, and glass bead blasted. It will receive transparent statuary finishes well. It finds use in many interior decorative applications, but when used in an exterior application it can be prone to dezincification if not cleaned and protected regularly (Figure 2.5).

Alloy C28000 got its unique name from its inventor, George Fredrick Muntz, a metal entrepreneur back in 1832 England. He was the proprietor of a firm called Muntz's Patent Metal Company and created the alloy for the cladding of ship hulls as a biocide to protect the wooden surfaces from marine growth that damaged the ships.

LEADED BRASSES

Leaded brasses are similar in many ways to unleaded brasses, but lead is added to the alloy to make it easier to machine. The lead introduces a property sometimes referred to as "free machining." As the cutting tool passes over the leaded alloy, small shards are expelled during the heat of cutting. In the case of Architectural Bronze, the lead allows for extruding and gives this alloy excellent machining

TABLE 2.8 List of leaded copper–zinc alloys.

Alloy	Zinc (%)	Lead (%)	Name	Color
C33000	34.5	0.5	Low-lead brass	Golden yellow
C35300	36	2.0	High-lead brass	Yellowish gold
C36000	34	3.0	Free-cutting brass	Reddish gold
C37700	38	2.0	Forging brass	Golden bronze
C38500	40	3.0	Architectural Bronze	Golden bronze

properties. Architectural Bronze is sometimes referred to as leaded Muntz Metal since the alloy makeup is similar to Muntz Metal, with 3% lead replacing some of the copper (Table 2.8).

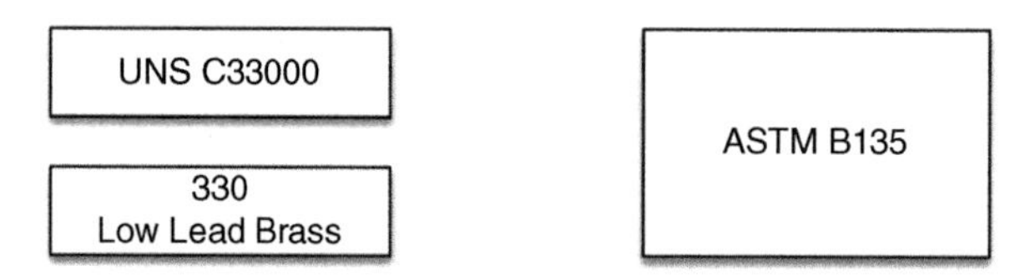

Alloying constituents added to Copper

Zn Pb

34.5% 0.5%

UNS designation	Temper	Tensile strength	Yield strength	Elongation % (2 in.–50 mm)	Rockwell Hardness B
C33000	H80	75 ksi (515 MPa)	60 ksi (415 MPa)	10	—

Available Forms

Tube

Alloy C33000 comes in tube forms. The tubes are hard drawn and the lead facilitates the drawing process. This alloy is not common in art and architecture. It is sometimes used as tubing for wiring in fixtures that require a golden color. The H80 temper designation stands for "hard drawn," which describes the temper produced as the tube is fabricated.

UNS C35300

353
High Lead Brass

ASTM B121/B121M
ASTM B453/B453M

Alloying constituents added to Copper

34.5% 2.0%

UNS designation	Temper	Tensile strength	Yield strength	Elongation % (2 in.–50 mm)	Rockwell Hardness B
C35300	H01	54 ksi (372 MPa)	40 ksi (276 MPa)	20	55
C35300	H02	60 ksi (413 MPa)	50 ksi (345 MPa)	38	70

Available Forms

Sheet
Plate
Rod

Alloy C35300 is used for stampings and building hardware for interior ornamentation. The added lead facilitates forming and machining. It also finds use in engraving. The 2% lead level allows the hard copper alloy to be machined very accurately.

UNS C36000

360
Free Cutting Brass

ASTM B16

Alloying constituents added to Copper

35.4% 3.1%

UNS designation	Temper	Tensile strength	Yield strength	Elongation % (2 in.–50 mm)	Rockwell Hardness B
C36000	H02	56 ksi (386 MPa)	45 ksi (310 MPa)	20	62
C36000	M30	49 ksi (338 MPa)	18 ksi (124 MPa)	50	—

Available Forms

Sheet
Plate
Rod
Extrusions
Bar

Alloy C36000 has good corrosion resistance and excellent machining characteristics. It is normally available in the H02 temper for flat material and M30 temper for hot extruded forms. Architecturally it is used as hardware and trim where machining is extensive. It has a pleasing yellow color.

UNS C37700

377
Forging Brass

ASTM B124
ASTM B283

Alloying constituents added to Copper

38% 2.0%

UNS designation	Temper	Tensile strength	Yield strength	Elongation % (2 in.–50 mm)	Rockwell Hardness B
C37700	M30	52 ksi (360 MPa)	20 ksi (140 MPa)	30	74

Available Forms

Hot drawn plate
Hot drawn rod
Hot drawn shapes

FIGURE 2.6 Alloy C37700 was used on the artwork for the Museum of the Bible.
Source: Artist Larry Kirkland.

Alloy C37700 is a forging and machining alloy. It is used where extensive shaping and machining is required. Its color is similar to alloy C28000 (Figure 2.6).

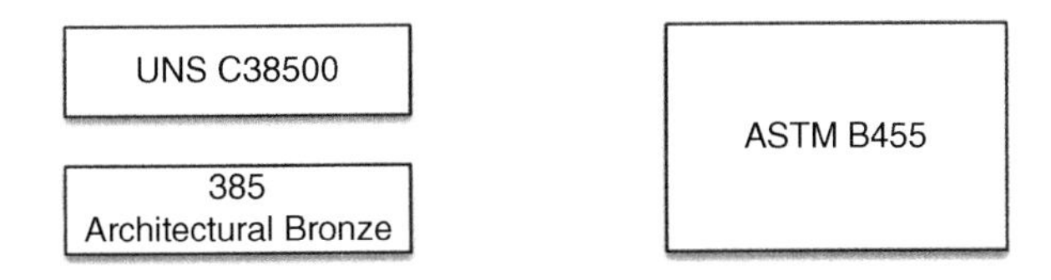

Alloying constituents added to Copper

Zn 40%

Pb 3.0%

UNS designation	Temper	Tensile strength	Yield strength	Elongation % (2 in.–50 mm)	Rockwell Hardness B
C38500	M30	60 ksi (413 MPa)	20 ksi (138 MPa)	30	65

FIGURE 2.7 Extruded plates of alloy C38500 (Architectural Bronze) used for the Los Angeles Police Department's Memorial to Fallen Officers by Gensler Architects.

Available Forms

Hot drawn shapes

Extrusions

As the name of alloy C38500 implies, Architectural Bronze is a common alloy used for architectural extruded forms. Brass handrails, door hardware, and extruded brackets and shapes are made from alloy C38500. Figure 2.7 shows a series of extruded Architectural Bronze plates. The color is a pleasing golden color that is similar to C28000; the alloy is essentially the leaded version of alloy C28000. It can be mirror polished, satin polished, and statuary finished. Alloy C38500 is very difficult to weld. The zinc alone makes it difficult, but the added lead in the alloying matrix precludes a good welded connection. Even if a weld can be achieved, it is usually pitted and porous, and the color will not match. Designing connections with joints and seams is recommended.

TIN BRASSES

Tin is added to improve the corrosion resistance of copper alloys. There are a number of tin brasses in industrial use. Alloys C44300, C44400, and C44500 are tin brasses that are also referred to as

Admiralty Metal. Admiralty Metal is sometimes called Arsenical Brass due to the small amounts of arsenic in the alloy mix. Today Admiralty Metal is used for tubing in heat exchange equipment, condensing equipment, and other applications where thermal transfer characteristics coupled with excellent corrosion resistance are required. These alloys are not commonly used in art or architecture.

The addition of a small amount of tin to alloy C28000, Muntz Metal, creates alloy C46400, Naval Brass. The small amount of tin greatly improves the corrosion resistance of this alloy. Adding 1% lead creates leaded Naval Brass. As the name suggests, these alloys are well suited for marine applications. They have good corrosion resistance and good strength. Naval Brass (alloy C46400) is now being used in art and architecture more frequently due to its higher corrosion resistance in chloride environments.

UNS designation	Zinc (%)	Tin (%)	Lead (%)	Name	Color
C46400	39	1	0.0	Naval Brass	Yellowish gold
C48200	38	1	1.0	Leaded Naval Brass	Yellowish gold

UNS C46400

464
Naval Brass

ASTM B21
ASTM B124
ASTM B283
ASTM B432

Alloying constituents added to Copper

Zn Pb Sn

39% –% 1.0%

UNS C48200

482
Leaded Naval Brass

ASTM B21
ASTM B124

Alloying constituents added to Copper

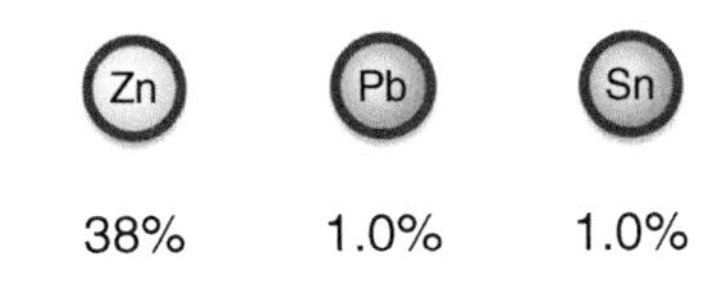

UNS designation	Temper	Tensile strength	Yield strength	Elongation % (2 in.–50 mm)	Rockwell Hardness B
C46400	H01	70 ksi (483 MPa)	58 ksi (400 MPa)	17	75
C48200	M30	63 ksi (435 MPa)	33 ksi (228 MPa)	34	60

Available Forms

C46400

- Sheet
- Plate
- Rod
- Bar
- Tube

C48200

- Fasteners
- Rods

Alloy C46400 has a color similar to alloy C28000 but better corrosion resistance. The added tin works to resist dezincification corrosion that C28000 is prone to in marine and industrial environments.

PHOSPHOR–BRONZE ALLOYS

Phosphor–bronze alloys contain a relatively large percentage of tin in their alloy constituents. They have been used on rare occasion in the architectural market. They have a less polished appearance and a ruddy color. The phosphor bronzes weather uniformly and the tin added to the copper improves corrosion resistance. These alloys weather well in seaside exposures. There are several phosphor–bronze alloys available. Phosphor bronzes are not considered common architectural alloys, but their corrosion resistance is good and the color is a pleasing reddish brown to golden brown.

UNS designation	Zinc (%)	Tin (%)	Phosphorus (%)	Lead (%)	Name	Color
C51000	0	5	0.2	0.0	Phosphor bronze A	Reddish brown
C54400	4	4	0.0	4.0	Phosphor bronze B2	Reddish brown

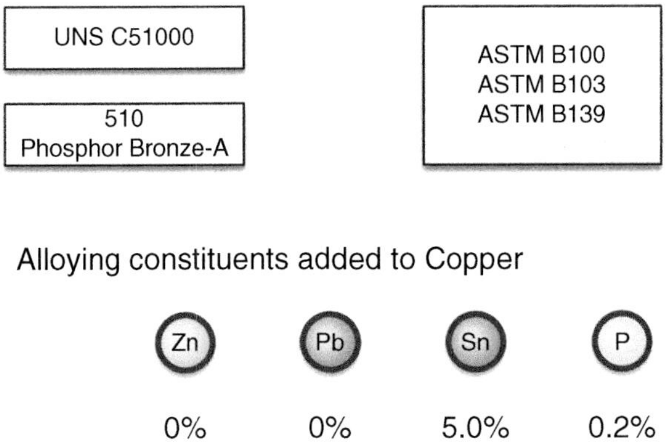

UNS C51000

510
Phosphor Bronze-A

ASTM B100
ASTM B103
ASTM B139

Alloying constituents added to Copper

Zn	Pb	Sn	P
0%	0%	5.0%	0.2%

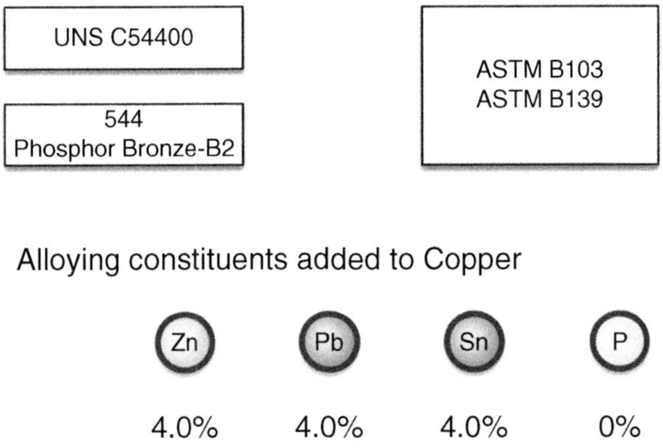

UNS C54400

544
Phosphor Bronze-B2

ASTM B103
ASTM B139

Alloying constituents added to Copper

Zn	Pb	Sn	P
4.0%	4.0%	4.0%	0%

UNS designation	Temper	Tensile strength	Yield strength	Elongation % (2 in.–50 mm)	Rockwell Hardness B
C51000	H01	49 ksi (338 MPa)	20 ksi (138 MPa)	24	—
C51000	H02	66 ksi (455 MPa)	54 ksi (373 MPa)	24	78
C54400	H02	58 ksi (400 MPa)	40 ksi (274 MPa)	24	68

These alloys are not commonly used in architectural or art. They oxidize to a ruddy surface color, do not polish out well, and can develop a patina. They are used when the design calls for a tough, gold-copper surface. They have good resiliency and good corrosion resistance. They can be used in marine environments where good chloride resistance is required. Due to their hardness and strength, they are often used as bearing plates. Alloy C54400 has good machineability and offers excellent wear resistance. Alloy C51000 has good strength and corrosion resistance but does not machine as well as C54400 due to the lack of lead.

ALUMINUM–BRONZE ALLOYS

Aluminum–bronze alloys are excellent marine alloys. They will not corrode in chloride exposures. They possess excellent resistance to corrosion in all normal environments. The aluminum–bronze alloys develop a thin, clear, aluminum-rich film over the surface that will keep the surface from developing a tarnish of copper oxide. The aluminum–bronze alloys are the most tarnish resistant of the copper alloys, and their appearance will remain a consistent yellow-gold color in most environmental exposures. They are not as easy to work as other copper alloys due to their hardness and are usually supplied with a cold rolled surface.

UNS designation	Al (%)	Tin (%)	Iron (%)	Nickel	Name	Color
C61000	8	0	0	0	Aluminum bronze	Yellowish gold
C61300	6.8	0.35	2.5	0	Aluminum bronze	Yellowish gold
C61400	7	0	2	0	Aluminum bronze	Yellowish gold
C61500	8	0	0	2	Aluminum bronze	Silver gold

UNS C61000

610
Aluminum Bronze

ASTM B169

Alloying constituents added to Copper

8.0% 0% 0%

UNS C61400

614
Aluminum Bronze

ASTM B150
ASTM B169
ASTM B171
ASTM B111

Alloying constituents added to Copper

7.0% 0% 2.0%

UNS C61500

615
Aluminum Bronze

ASTM B150
ASTM B169

Alloying constituents added to Copper

Al	Sn	Fe	Ni
7.0%	0%	0%	2.0%

UNS designation	Temper	Tensile strength	Yield strength	Elongation % (2 in.–50 mm)	Rockwell Hardness B
C61000	O60	52 ksi (359 MPa)	17 ksi (117 MPa)	45	60
C61300	H50	75 ksi (517 MPa)	45 ksi (310 MPa)	30	—
C61400	O60	80 ksi (552 MPa)	40 ksi (276 MPa)	40	85
C61500	O50	85 ksi (586 MPa)	50 ksi (345 MPa)	36	70

Available Forms

Alloys C61000 and C61500 are typically provided in sheet.

Alloy C61300 is used as hot rolled and extruded shapes.

Alloy C61400 is used for fasteners and tube shapes.

The aluminum–bronze alloys have good strength and excellent corrosion resistance.

They are not as common in architectural use as many of the other bronze and brass alloys; however, alloy C61500 is finding more use in exterior wall and roof applications. Alloy C61500 has excellent tarnish resistance and weathers to a matte golden color. It can be prepatinated. The patina colors are brownish golden tones. These alloys do not receive a mechanical finish well due to their hardness even in annealed tempers. Alloy C61500 in a half-hard temper has a tensile strength of 105 ksi (724 MPa). For architectural uses, the annealed temper is preferred for forming and piercing.

Alloys C61000, C61300, and C61400 are not as common in architectural uses. C61000 is used as ornamental trim, while alloys C61300 and C61400 are used as fasteners and in industrial equipment as well as in sheathing for marine exposures. They are more expensive than copper or brass, thus they have found limited use (Figure 2.8).

FIGURE 2.8 Alloy C61000 form made from thin sheet.

SILICON–BRONZE ALLOYS

Alloy C65500 is the wrought form of silicon bronze. Most art and architectural applications of silicon bronze use the cast form of the silicon–bronze alloys.

UNS C65500

655
Silicon Bronze
High Silicon Bronze A

ASTM B96
ASTM B98
ASTM B100
ASTM B124
ASTM B315

Alloying constituents added to Copper

1.5% max 3.0% 1.0% max

UNS designation	Temper	Tensile strength	Yield strength	Elongation % (2 in.–50 mm)	Rockwell Hardness B
C65500	H01	68 ksi (469 MPa)	35 ksi (241 MPa)	30	75
C65500	H02	78 ksi (538 MPa)	45 ksi (310 MPa)	17	87

Available Forms

Sheet
Plate
Rod
Bar
Tubes
Wire

In art and architecture, alloy C65500 is generally incorporated into sculpture in a plate or rod form that matches the cast silicon bronze form. In its mill, clean, unoxidized state, the color of alloy C65500 is very close to commercially pure copper alloys, with a tone more brown than the salmon red color of copper. The silicon improves machinability of this alloy and increases its strength and hardness.

This alloy is available in sheets as thin as 0.6 mm (0.025 in.) and plates up to 19 mm (0.75 in.) thickness.

COPPER–NICKEL AND NICKEL–SILVER ALLOYS

These attractive alloys are more expensive than other alloys of copper due to the addition of nickel. Nickel, used extensively in the production of austenitic stainless steels and as plating over brass and steel, has a value that fluctuates more radically than other metals in architecture. Its price per unit weight is several times that of copper. This places a constraint on the metal as an architectural metal and the availability of these alloys to the art and architecture market is limited both in form and production. When working with these alloys, verifying the source and constraints that exist is recommended. Specific widths, thicknesses, and quantities are not always readily available.

Copper–nickel alloys have excellent corrosion resistance and will perform better than most of the other alloys of copper in terms of corrosion in marine exposures. Copper–nickel alloys have very high strength and can be both cold and hot worked. The hardness increases rapidly when cold worked.

The nickel–silver alloys of copper have an attractive color of golden silver (Table 2.9). They can be polished and highlighted with oxides. There is no actual silver in these alloys; the name is derived from the original mystery around the metal and color.

TABLE 2.9 List of nickel–silver alloys used in architecture.

UNS designation	Zinc (%)	Nickel (%)	Manganese (%)	Iron (%)	Name	Color
C70600	1	10	10	1	Copper nickel 10	Reddish silver
C71500	1	30	1	1	Copper nickel 30	Silver
C75200	17	18	0	0	Nickel silver 65-18	Golden silver
C77000	27	18	0	0	Nickel silver 55-18	Golden silver

These alloys were in wide use in the late 1800s and early 1900s as the "white metal" before stainless steel and aluminum were introduced into the world of art and architecture. This pleasing golden silver metal was incorporated into the ornamentation of the art deco age where it offered a beautiful contrast to polished brasses. Before that it was used for ornamental dining wear, custom boxes, and jewelry.

The metal has its roots in China: it is said to have been produced in China during the Qing dynasty sometime in late seventeenth century. Germany imported this copper–nickel alloy in the early 1700s from China, where it was called *paktong*, or "white copper." The metal was considered an alternative to silver, with a similar weight, feel, and color. It held a mystique when it first arrived in Europe and found use as a precious metal for ornamentation. The Chinese are said to have smelted both copper ore and nickel ore separately. Later, using special techniques of sublimation, they added zinc to the mixture in order to make it more malleable.[2] The technique the Chinese artisans used to add the zinc was remarkable in the sense that it was different from the normal approaches used to alloy metals in those times, which used a vapor sublimation technique to add zinc to heated copper sheets and then added the coated copper sheets to be melted with the nickel.

In 1823 Germany sponsored a competition to improve the quality and color of this alloy. The idea was to make it more like silver. The competition was won by the Hemminger brothers of Berlin and Earnst August Grither of Sheneeberg, who created what we know today as the nickel–silver alloys C75200 and C77000.

Nickel silver has been known by various names over the years: *paktong*, German Silver, new silver, alpaca, Paris Metal, Maillechort (derived from the names of two French metallurgists, Maillot and Chorier, who eventually developed the French version of the alloy in 1820), and even Dairy Bronze for the cast version of the metal (alloy C97600), which is used to cast fittings and valves for the dairy industry.

Today it is not as commonly used in art and architecture, although it possesses a beautiful color (Figure 2.9). Its cost and availability have placed a limit on its use. Mill sources do not produce the metal on a regular basis, so supplies and available dimensions are limited.

[2]Keith Pinn, *Paktong: The Chinese Alloy in Europe 1680–1820* (London: Antique Collectors Club, 2006), 33–35.

FIGURE 2.9 Interior walls made of alloy C75200.

UNS C70600

706
Copper Nickel 10

ASTM B122
ASTM B151
ASTM B171
ASTM B111

Alloying constituents added to Copper

10% 1.0% 1.0% 1.0%

UNS C71500

715
Copper Nickel 30

ASTM B122
ASTM B151
ASTM B171
ASTM B111

Alloying constituents added to Copper

30% 1.0% 1.0% 1.0%

UNS designation	Temper	Tensile strength	Yield strength	Elongation % (2 in.–50 mm)	Rockwell Hardness B
C70600	H01	60 ksi (415 MPa)	48 ksi (330 MPa)	20	58
C71500	O61	55 ksi (380 MPa)	18 ksi (125 MPa)	36	40

Available Forms

Sheet

Plate

Tube

Rod (alloy C71500 only)

UNS C75200

752
Nickel Silver 65-18

ASTM B122
ASTM B151
ASTM B206

Alloying constituents added to Copper

18% 17%

UNS C77000

770
Nickel Silver 55-18

ASTM B122
ASTM B151
ASTM B206

Alloying constituents added to Copper

18% 27%

UNS designation	Temper	Tensile Strength	Yield Strength	Elongation % (2 in.–50 mm)	Rockwell Hardness B
C75200	H01	65 ksi (450 MPa)	50 ksi (345 MPa)	30	73
C77000	OSO35	60 ksi (415 MPa)	27 ksi (185 MPa)	40	55
C77000	H04	100 ksi (690 MPa)	85 ksi (585 MPa)	3	91

Available Forms

Plate

Sheet

Rod

Wire

UNS C79800

798
Leaded Nickel Silver

ASTM B122
ASTM B151
ASTM B206

Alloying constituents added to Copper

9–11% 36–42% 1.5–2.5% 0.25 max

Available Forms

Extrusion

The nickel–silver alloys C75200 and C77000 possess different colors due to differences in zinc content. It is important to note the strength differences of the different tempers. Alloy C77000 can be difficult to work in the cold rolled temper. The OS35 temper controls grain size by annealing, which will allow a drop in strength for further cold work (Figure 2.10).

FIGURE 2.10 Art cutout using alloy C75200.

These alloys get a further boost in strength and hardness from the zinc addition, while nickel provides the silvery color. Nickel silvers will eventually weather to produce a patina of greenish brown. Their excellent corrosion resistance in many environmental exposures means they will resist the development of a patina, with heavy coastal exposures and severe industrial exposures being the exception.

CAST ALLOYS

There are a number of copper alloys used for casting. The casting of copper alloys is one of the oldest forms of metalworking, going back centuries to the beginning of civilization. High-purity copper is not cast because of the difficulty working the metal. Copper absorbs gases when in the molten form, causing it to swell; later as the metal cools the gases are freed, creating porosity.

Art foundries around the world are expert at casting copper alloys for sculpture and techniques have been refined and perfected for centuries. There are numerous commercial foundries that cast copper alloys for architectural hardware and fittings. Cast alloys can have identical appearance and color to the wrought alloys, and they can be polished and chemically enhanced with custom oxides and patinas.

Most copper alloy castings, in particular cast sculpture, are often called "bronze." The term "bronze" is more a descriptive term for color rather than for alloy. True bronze is an alloy of copper and tin, but over the centuries the term "bronze" has been used to describe a number of alloys that have no tin in them.

In different parts of the world, different copper alloys are used for casting. For statuary casting in the United States, two alloys, sometimes referred to by the trade name Herculoy[3] or Everdur™, are in common use. These are both versions of silicon bronze with differences in alloying elements. Herculoy is a trade name for the silicon–bronze alloy C87610 and is produced by Revere Copper, while Everdur was developed by the DuPont company in the early 1920s. DuPont wanted a metal alloy that had the strength of steel but was corrosion resistant. Everdur was patented and trademarked and the DuPont Everdur Company was established in Delaware.

In Europe, a copper alloy consisting of tin, zinc, and lead is used for casting. This alloy is called "Leaded Gunmetal" and sometimes goes by the initials LG2 or is called the 85-5-5-5 alloy. It closely resembles alloy C38600. This alloy was originally used to cast cannons and is considered by some foundries a superior alloy for casting sculpture because of the ease of finishing and the variety of patinas that can be developed on the surface. The use of alloy C83600 is being slowly phased out because of the lead used in the alloying mix and the hazards that lead, in quantity, poses to the workers in the foundry and concern over its disposal. In some parts of Asia traditional bronze is still used for casting. The alloy is 90% copper and 10% tin and closely resembles C90500. Another alloy composed of 90% copper, 5% tin, and 5% zinc is a common cast alloy in India.

The C87300 silicon bronze alloy is often considered for cast statuary in the Americas and is increasingly being considered in Europe. Asia is also beginning to adopt the C87300 alloy for casting bronze art sculpture. The reasons that the silicon–bronze alloys are being used more often lie in the availability of the metal, the casting process, and the ease of pouring the metal and arriving at good detail within the cast.

There are other copper alloys that have been used for casting. These alloys are used to produce specific properties or specific coloration. Leaded red brass, for example, has been used to cast intricate forms with elaborate detail (Tables 2.10 and 2.11).

TABLE 2.10 Common cast tempers.

Cast tempers	Previous description
M01	As sand cast
M02	As centrifugal cast
M03	As plaster cast
M04	As pressure die cast
M05	As permanent mold cast
M06	As investment cast
M07	As continuously cast

[3]Herculoy® is a registered trademark of Revere Copper Products of Rome, New York. It was registered in 1930.

TABLE 2.11 Common cast alloys.

UNS designation	Name	Shrinkage (%)	Liquidus (°C)	Castability rating	Fluidity rating	Use
C83600	Leaded red brass	5.7	1010	2	6	Art work
C84400	Semi–red leaded brass	2.0	980	2	6	Extensive machining
C85200	Leaded yellow brass	1.5–1.8	940	4	4	Good machining Pleasing color
C85400	Leaded yellow brass	1.5–1.8	940	4	3	Plaques, ornamental castings
C85800	Yellow brass	2.0	925	4	3	Die casts
C86300	Manganese bronze	2.3	920	5	2	Hardware
C87200, C87300	Silicon bronze, Everdur	1.8–2.2	916	5	3	Art work
C87500	Silicon brass Tombasil	1.9	915	4	1	Hardware
C87600	Low-zinc silicon brass	1.8–2.0	971	4	1	Art work
C87610	Silicon bronze	1.8–2.2	971	4	1	Art work, general
C90300	Tin bronze	1.5–1.8	980	3	6	Gears, bushings
C97300	Leaded nickel silver	2.0	1040	8	7	Ornamental castings
C97600	Dairy Metal	2.0	1145	8	7	Hardware
C97800	Leaded nickel bronze	1.6	1180	8	7	Ornamental castings, musical instruments
C99700	White manganese brass	2.0	902	3	1	Bearings

For castability and fluidity, 1 is the highest rating.
Source: *ASM Specialty Handbook: Copper and Copper Alloys* (Materials Park, OH: ASM, 2001), 92.

RED BRASSES

UNS C83600

Leaded Red Brass
Leaded Gunmetal
LG2
85-5-5-5
European Bronze
Ounce Metal

ASTM B30
ASTM B62
ASTM B271
ASTM B505
ASTM B584

Alloying constituents added to Copper

4–6% 4–6% 4–6% 0.3% max 1% max

Aluminum, antimony, phosphorus, silicon, and sulfur may be present in small amounts.

UNS designation	Temper	Tensile strength	Yield strength	Elongation (%)	Hardness (Brinell)
C83600	M01	30 ksi (205 MPa)	14 ksi (95 MPa)	30	60

Alloy 83600 is not as strong as the silicon–bronze alloys, but it has good fluidity. The hardness allows for easier finishing of the surface. This alloy has been extensively used throughout Europe over the years, and many European foundries still use it and consider it a superior casting alloy for art and sculpture. The color is golden with a reddish tone. It takes patination well, particularly with the darkened statuary finishes.

The lead content gives good machining properties, but the amount of lead in the alloy limits the use of C83600 in art castings (Figure 2.11).

UNS C84400

Leaded Semi-Red Brass
Valve Metal

ASTM B30
ASTM B62
ASTM B271
ASTM B505
ASTM B584

Alloying constituents added to Copper

2–3.5% 7–10% 6–8% 0.4% max 1% max

Aluminum, sulfur, phosphorus, and silicon may be present in small amounts.

FIGURE 2.11 The use of alloy C83600 for the cast doors at the Kansas City Life Insurance building.

UNS designation	Temper	Tensile strength	Yield strength	Elongation (%)	Hardness (Brinell)
C84400	M01	34 ksi (234 MPa)	15 ksi (103 MPa)	26	55

Alloy C84400 has a little more strength and the added lead makes it more fluid and softer. It is not an alloy commonly used in art and architecture. The lead content (as high as 8%) makes this alloy less desirable to work with in the foundry. Alloy C84400 is occasionally cast as ornamental hardware.

YELLOW BRASSES

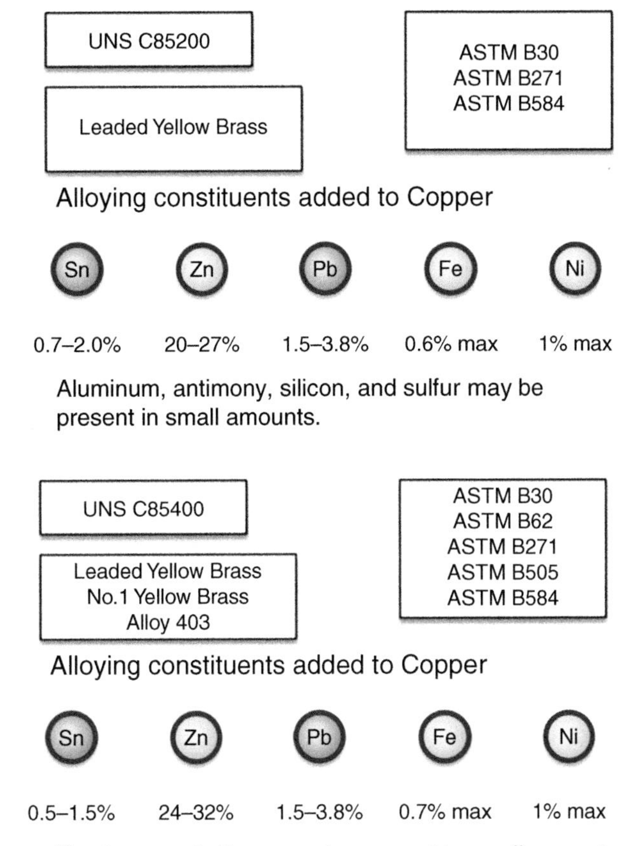

UNS designation	Temper	Tensile strength	Yield strength	Elongation (%)	Hardness (Brinell)
C85200	M01	38 ksi (260 MPa)	13 ksi (90 MPa)	35	45
C85400	M01	38 ksi (262 MPa)	14 ksi (95 MPa)	35	50

The C85200 and C85400 cast alloys are very similar, and both are used in art and architecture. C85200 is used as an ornamental brass casting, similarly to yellow brass. C85400 is used in plaques and ornamental fixtures where the yellow color is desirable.

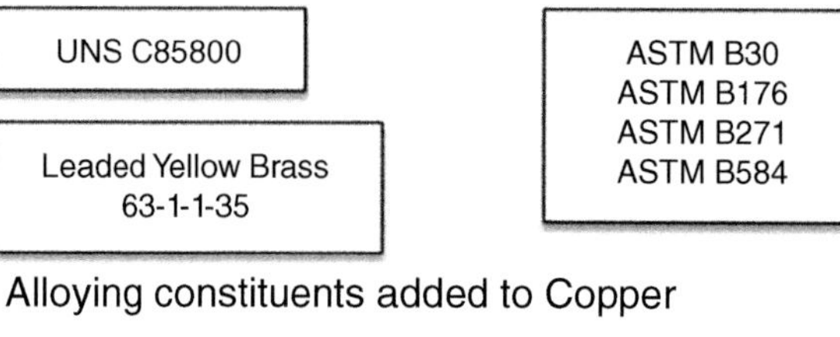

Alloying constituents added to Copper

1.5% max 31–41% 1.5% max 0.5% max

UNS designation	Temper	Tensile strength	Yield strength	Elongation (%)	Hardness (Brinell)
C85800	M01	55 ksi (380 MPa)	30 ksi (205 MPa)	15	102

Alloy C85800 is a die cast alloy. Hardware and small ornamental forms are made from die castings of C85800. The castings have a quintessential brass look with high strength and hardness.

Alloying constituents added to Copper

0.2% max 22–28% 2.5–5% 0.2% max 2–4% 3–7.5%

UNS designation	Temper	Tensile strength	Yield strength	Elongation (%)	Hardness (Brinell)
C86300	M01	119 ksi (820 MPa)	67 ksi (460 MPa)	18	225

Alloy C86300 is extra strong. It is not a common architectural alloy and is instead used for gears and bearing plates. Alloy C86300 does not machine well and does not perform well in marine exposures.

SILICON–BRONZE ALLOYS

Silicon–bronze alloys melt at 971 °C (1870 °F)—known as the "liquidus point" and begins to return to a solid state—the "solidus"—at 821 °C (1510 °F). Their low melting point results in energy economy in the casting process and they are therefore favored by foundries. Silicon bronze has a high fluidity rating: in other words, it flows nicely into detailed casting.

"Fluidity" is the term given to the ability of a molten metal to travel before solidifying. This differs from the castability of a metal, which is the measure of both a metal's shrinkage characteristic once it begins to solidify and a metal's freezing range, or the range of temperatures at which solidification starts to occur. For silicon bronze, shrinkage can be expected in the 1.8–2% range. This is critical for the design of the mold being cast into. The fluidity of silicon–bronze alloys provides highly detailed casting in the mold, but they are subject to what is called "directional solidification," a condition in which shrinkage cracks can occur. Proper mold design and correct gating and riser placement is critical to keep metal available as shrinkage occurs.

Silicon bronze is in the middle range with respect to castability. Mold design must be such that there is good feed to all areas of molten metal with good exothermic protection to prevent premature solidification. Keeping dross out of the casting is also important for a good surface and a sound cast part. You want the molten metal to enter the mold at a controlled rate and keep turbulence down, otherwise dross will develop and result in a porous casting. These alloys have little to no lead.

Following are the more common silicon–bronze alloys used in art and architecture. Some are also referred to as silicon brasses due to their high zinc content. Other silicon–bronze alloys have been created for industry, but those that are listed are used in the art sculpture and architectural casting industry. These alloys are high in copper content, containing at least 90% copper.

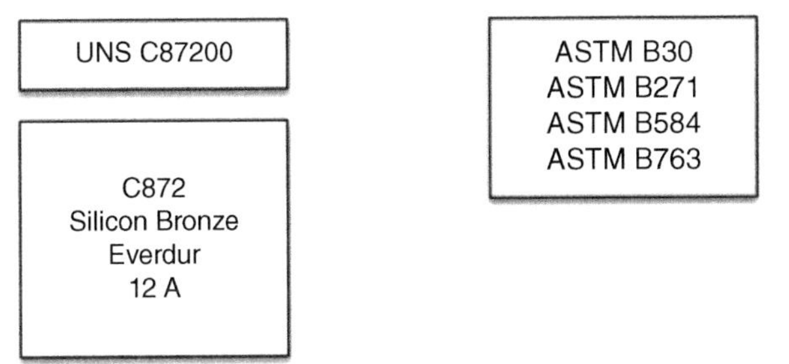

1–5% 5% max 1.5% max 0.5% max 2.5% max 1% max 1.5% max

UNS C87300

C873
Silicon Bronze
95-1-4
Everdur
Herculor
Tombasil
Grade A

ASTM B30
ASTM B271
ASTM B585
ASTM B763

Alloying constituents added to Copper

3–4.5% 0.25% max 0.8–1.5% 0.09% max 0.2% max

UNS designation	Temper	Tensile strength	Yield strength	Elongation (%)	Hardness (Brinell)
C87200	M01	45 ksi (310 MPa)	18 ksi (124 MPa)	20	35
C87300	M01	45 ksi (310 MPa)	18 ksi (124 MPa)	30	35

Alloys C87200 and C87300 are general-purpose silicon–bronze alloys used for casting. They are interchangeable and basically the same. They have been called Everdur, Herculoy, silicon bronze, grade A silicon bronze, and other names through the years as foundries in the United States embraced these alloys. They are the most common cast bronze alloys used in statuary in the United States and are becoming more popular in Europe and Asia. Everdur was initially developed back in the 1920s by DuPont for use in chemical facilities that needed an antisparking metal.

One drawback of the silicon–bronze alloys is the difficulty of achieving a broad range of patina colors. The silicon content works against the patina process and the high copper content makes it highly resistant to corrosion. Transition oxide layers are needed to achieve good variations in color. The closest wrought alloy is C65500, a silicon bronze that also goes by the name Everdur.

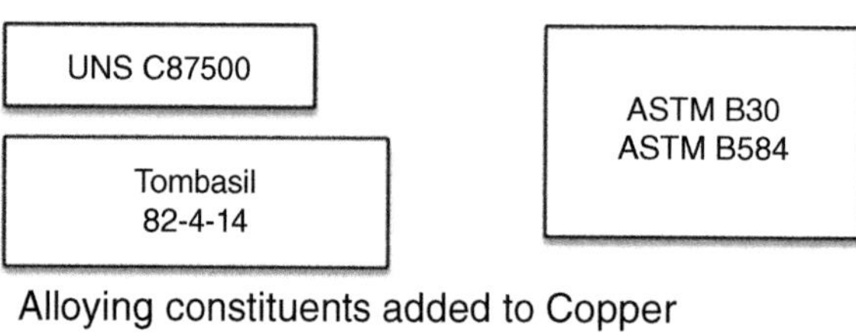

Alloying constituents added to Copper

12–16% 3–5% 0.5% max 0.5% max

UNS designation	Temper	Tensile strength	Yield strength	Elongation (%)	Hardness (Brinell)
C87500	M01	67 ksi (460 MPa)	30 ksi (205 MPa)	21	134

Alloy C87500 is a sand cast alloy of what is also known as Tombasil.. It is very strong and has excellent hardness. It is also known as silicon brass. Alloy C87800 is the die cast version of the alloy and contains small amounts of tin, manganese, magnesium, and iron.

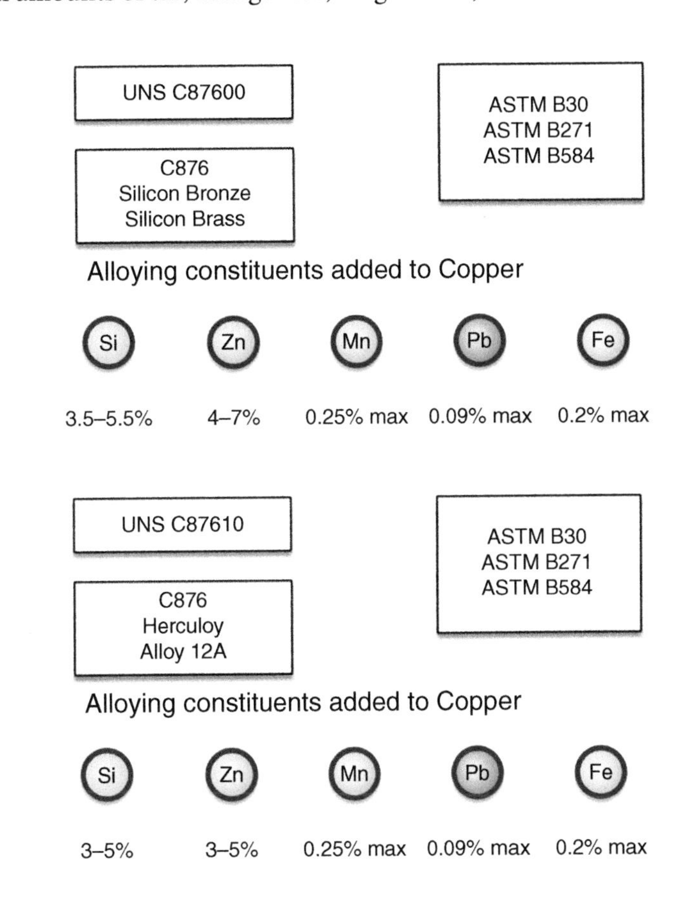

UNS designation	Temper	Tensile strength	Yield strength	Elongation (%)	Rockwell Hardness B
C87600	M01	66 ksi (455 MPa)	32 ksi (220 MPa)	20	76
C87610	M01	55 ksi (380 MPa)	25 ksi (172 MPa)	30	76

Alloys C87600 and C87610 are two similar alloys used in casting. Both are high in silicon and zinc. Alloy C87600 has the greater amount of zinc and thus is harder and stronger. A range of patinas can be induced in these alloys. They are used in art casting. The higher silicon content makes casting easier by improving fluidity and lowering the liquidus temperature. Strength improves but ductility is sacrificed somewhat.

A TIN–BRONZE ALLOY

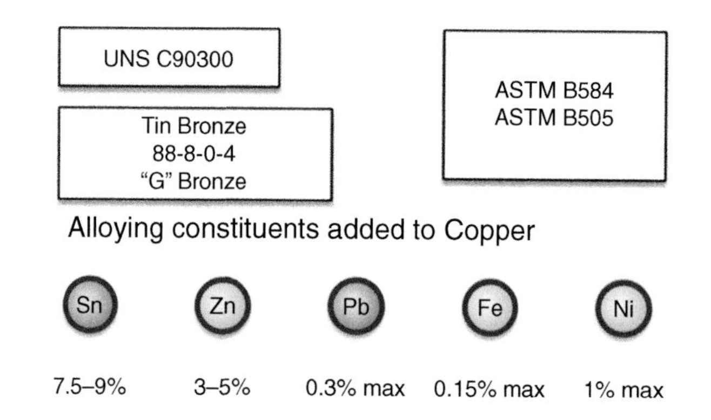

UNS designation	Temper	Tensile strength	Yield strength	Elongation (%)	Hardness (Brinell)
C90300	M01	45 ksi (310 MPa)	21 ksi (145 MPa)	30	70

Alloy C90300 has excellent strength. It is not an alloy commonly used in art and architecture. It is used when wear friction and wear resistance are required.

NICKEL–SILVER ALLOYS

UNS C97300

12% Nickel Silver
Leaded Nickel Brass
Alloy 10A

ASTM B271
ASTM B505
ASTM B584

Alloying constituents added to Copper

Ni	Zn	Sn	Fe	Pb
11–14%	17–25%	1.5–3%	1.5% max	8–11%

UNS C97600

Dairy Metal
20% Nickel Silver
Alloy 11A

ASTM B271
ASTM B505
ASTM B584

Alloying constituents added to Copper

Ni	Zn	Sn	Fe	Pb
19–21.5%	3–9%	3.5–4%	1.5% max	3–5%

UNS C97800

25% Nickel Silver
Alloy 11B

ASTM B271
ASTM B505
ASTM B584

Alloying constituents added to Copper

Ni	Zn	Sn	Fe	Pb
24–27%	1–4%	4–5.5%	1.5% max	1–2.5%

UNS designation	Temper	Tensile strength	Yield strength	Elongation (%)	Hardness (Brinell)
C97300	M01	34 ksi (230 MPa)	16 ksi (110 MPa)	9	55
C97600	M01	45 ksi (310 MPa)	20 ksi (140 MPa)	11	80
C97800	M01	54 ksi (370 MPa)	25 ksi (170 MPa)	10	130

The cast nickel silvers have a relatively high lead content, which aids in machining. These alloys are sometimes used in ornamental castings and ornamental plaques. Corrosion-resistant hardware that requires toughness is also cast from these alloys, and they have a pleasing color. These alloys are soldered and brazed but the lead makes them unsuitable for welding.

A MANGANESE–BRONZE ALLOY

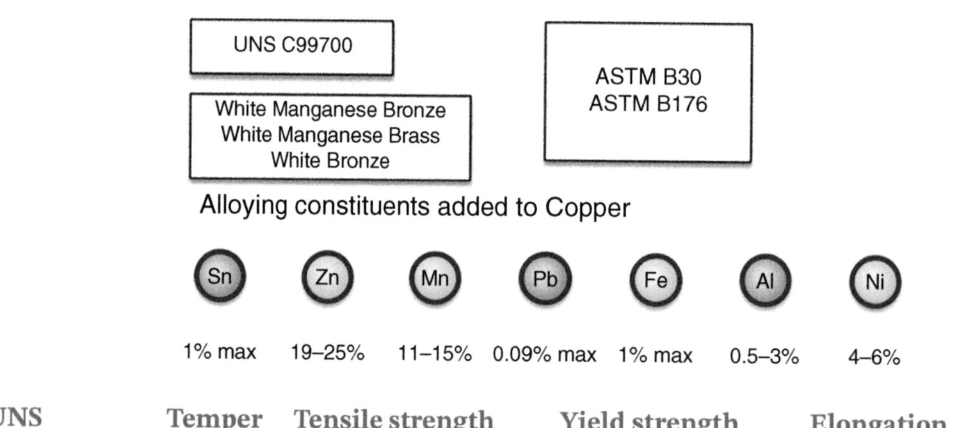

UNS designation	Temper	Tensile strength	Yield strength	Elongation (%)	Hardness (Brinell)
C99700	M01	55 ksi (380 MPa)	25 ksi (172 MPa)	25	110

Alloy C99700 has good casting characteristics, toughness, and strength. It has a silvery gold color and is used in ornamental castings. It machines well.

The preceding by no means covers the entire gamut of copper alloys. The list of alloys is extensive and continues to grow as new alloys are created for specific properties to be used in industry. The tempers that are listed and their approximate strengths are also only a partial list.

It's important to remember that just because an alloy formulation exists does not mean that it is readily available. As with all industries, there is a matter of scale. For reasons of practicality, not every alloy is stocked—in reality, very few are. A Mill producer of a copper alloy casts several tons of an alloy in a single cast and therefore either bundles orders for the particular alloy to be cast or requires the purchase of an entire cast lot.

Tempering also is a process carried out at a large scale. It is performed at the Mill, and like casting requires a large quantity for a specific temper.

CHAPTER 3

Surface Finishes

Patinas . . . at their best they create a bit of magic or poetry that sings off the form.

Rungwee Kingdon

INTRODUCTION

The surfacing of copper and copper alloys is an art form in itself. The color and appearance of newly rolled copper plate and sheet are unmistakable. Their beautiful color and gleaming luster has captivated us through the centuries. The process the Mill undertakes to produce the finish on copper alloys produced for industrial use demonstrates care for the intrinsic beauty of this metal.

The Mill produces a clean, smooth surface on copper and copper alloys. On thick plates the surface can be rough and uneven with surface inclusions, but thin sheets and plates, extrusions, rods, wire, and foil are universally provided clean and free of oxides by the Mill producer. The beauty and luster of the metal demands that the metal be handled with care as it is formed and shaped into the final product.

Newly cast copper alloys that have been blasted clean also exhibit a special beauty. It seems almost a shame to develop a patina over this beautiful surface. However, once the patinuer (an artisan who applies the finish) begins, the metal comes alive with a color that possesses the essence of what we consider to be the metal. The detail of the art takes form as if a fog were receding. Colors enhance the contrasting shadows, and as the patina develops under the heat of the torch a new richness, natural as a polished stone, takes shape. Some patinas are transparent, allowing the natural beauty of the metal to reflect through and giving depth and contrast.

This is the beauty of copper and copper alloys. They have a natural attractiveness that becomes more enchanting as the surface interacts with elements most other materials abhor. They capture these elements using their natural tendency to arrive at a stability afforded by copper's thermodynamic character. Few materials will combine with other elements so readily and with such beauty as copper and copper alloys.

If not protected, the newly exposed copper alloy surface will oxidize rapidly, eventually forming a darkened tarnish over the freshly cleaned surface. Hot, humid environments see the most rapid change, sometimes within hours. High humidity and moisture will cause the surface to react and form a layer of oxide. Thus, it is important to keep the metal dry and protected while in storage and until it is ready to be installed.

The aluminum–bronze alloys are an exception. They resist tarnish by rapidly developing a thin, clear layer of aluminum oxide on the surface. They maintain their golden yellow color longer than the other copper alloys in all conditions of exposure.

MILL SURFACES

Cold rolled copper products—including sheet, coils, rods, bars, and some plates—are produced with an "as fabricated" finish. For copper alloys this finish is usually smooth, clear, and reflective. The steel rolls used to reduce the thickness and impart the temper are clean and smooth. In all cases and with any of the alloys you should expect a smooth, oxide-free surface when cold rolling passes are applied to the forms. Even tubing produced at the Mill is clean and free of mars and scratches. The copper conduit tubing we often see may have scratches and marks, but this is from subsequent handling at material supply houses. The same general statement is applicable to the rod and bar forms of copper alloys.

At the Mill source a fine oil containing benzotriazole is usually applied once the finish has been induced onto the clean alloy surface. This imparts a level of tarnish protection while the metal is stored and handled in the semi-raw form of sheets, plates, and bars. This protection will suffice for a short duration if the metal is stored in a dry protected area.

An oxide may exist on the surface of heavy alloy plates. Figure 3.1 shows a hot rolled alloy plate that is flat and smooth but with a layer of adherent oxide on the surface. The thick copper plate in the left-hand image has more irregular scale on the surface than the plate on the right. In the left-hand image you can see the lines the rollers left behind as the 50-millimeter copper alloy plate was hot rolled at the Mill. The plate in the right-hand image is made of alloy C48200. The oxide is smooth across the surface of these narrow plates. Most likely a cold pass was performed to flatten the plates.

The Mill surface applied to copper is a cold rolled finish. It has good reflectivity but lacks the refinement of an applied mechanical surface. The surface texture is smooth and often possesses some directionality. The finish from one Mill source is by its nature different than the finish from another Mill source because a finish is defined by the smoothness and quality of the steel roll surface used during the final pass of the cold rolling process. Different mills will have different levels of polish on these cold rolls.

FIGURE 3.1 Heavy oxidation on a thick copper alloy plate.

This would also be the case for natural-finish copper alloys of different thicknesses, since the metal was created at different times and under different pressures and temper passes. The grains will be different, as the thicker material has not undergone the same extent of cold passes as the thinner metal.

The brass and bronze copper alloys are also provided with a cold rolled surface for thicknesses less than 5 mm. For thicknesses greater than 5 mm, the surface can be a hot rolled surface. In copper alloys the hot rolled surface may have some pitting and a surface grain appearance that is rougher than the cold rolled surface.

Typically copper alloys are finished further at a polishing facility or the fabrication facility. For this step the finish needs to be clean and smooth, and since few architectural or art features require very thick sheets or plates the finish almost always uses a cold rolled pass to arrive at a smooth surface for further polishing.

MECHANICAL FINISHES AND TEMPORARY PROTECTION

Mechanical finishes such as satin polishing, mirror polishing, and on occasion, glass bead blasting, are often used on copper and copper alloys when the design seeks a more refined surface. In addition, embossing and custom hammering are macrosurface deformations used on some of the softer alloys to add stiffness and to enhance appearance. These mechanical finishes demand special care due to the reactivity of the fresh, exposed copper alloy surface. This oxide-free surface will rapidly fingerprint and tarnish. Protective plastic films can be applied to the surface after the mechanical finish has been created; however, these plastic films offer only short-term protection from oxidation. Air, trapped under the film or entering through spaces on the edges, will darken the clean metal surface. Paper interleave saturated with antitarnishing solutions can provide temporary protection of the copper alloy surface. Both plastic film and saturated paper can leave residues that can interfere with additional chemical oxidation processes that may be planned for the metal surface.

Natural Color

The natural color that many copper alloys possess can be attractive and appealing, which is one of the reasons why over the centuries coins have been and still are made from copper and copper alloys. The bright color of newly minted coins reflects the sense of value associated with the metal and its color.

Brightly polished brass can be found in the lobbies of buildings housing institutions of power, wealth, or elegance. Elevator doors, entryways, handrails, and column covers made of copper alloys and polished to produce a highly reflective golden appearance convey a sense of elegance few other materials can match. The interiors of churches, synagogues, and mosques are adorned with ornamentation made from copper alloys that have been brightly polished to give a golden reflection, while many of their exteriors are clad in natural aging copper roofing.

In the early part of the twentieth century, nickel silver was used to provide a silvery tone and set alongside brightly polished brass (Figure 3.2). At the time these stunning art deco forms represented the modern age—and a sense of opulence that extends to today.

Capturing and maintaining the brightly polished surfaces of copper alloys is no easy task. Methods have been employed over the years that try to eliminate the powerful drive of the single valence electron of copper to combine with oxygen, the combination that results in the tarnish on the surface of copper alloys. This is not an issue, however, with the aluminum–bronze alloys. The aluminum bronzes C61000 and C61500 are the most corrosive resistant of the copper alloys and they resist oxidation by developing an aluminum-rich, clear oxide on the surface. In urban, coastal, or rural environments these alloys will maintain a consistent appearance and color over years of exposure.

FIGURE 3.2 Art deco nickel silver.

FIGURE 3.3 The aluminum–bronze alloy TECU Gold, produced by KME of Germany, was used for the information center at the Münster 07 sculpture exhibition.
Source: The structure was designed by büro modularbeat, Münster, Germany.

They are typically used in their bright, cold rolled surface, with no secondary finishing or polishing beyond what is accomplished from the condition of the roll used to develop the finish. Figure 3.3 shows a fascinating example of TECU® Gold, an aluminum–bronze alloy produced by KME. This lightweight pavilion reflects a golden yellow tone.

Constant polishing and cleaning are necessary to keep most copper alloys appearing bright and golden. They can be coated and waxed to reduce the frequency of the cleaning regimen, at least for works used in the interior of a structure, where the environment is more stable. One of the most effective lacquers in use over the last several decades is Incralac®, a product developed under the guidance of the International Copper Research Association (INCRA). This thermoplastic copolymer of acrylic contains benzotriazole, an effective tarnish inhibitor. Chapter 8 discusses the makeup and effectiveness of Incralac. When applied correctly to copper and copper alloys it

FIGURE 3.4 Alloy C28000, polished and coated with clear Incralac, in use at three building entryways.

gives years of protection, allowing the surface to resist oxidation. Figure 3.4 shows three examples of polished Muntz Metal (alloy C28000) coated with Incralac in use on the exterior entryways of various buildings.

Exterior roofing and thin copper wall cladding are rarely coated with protective clear coatings, even if given an artificially induced patina. It is better to allow them to react and change with the environment than to undertake the expensive and somewhat futile endeavor of maintaining their clear coatings. The de Young museum in Golden Gate Park, San Francisco, has walls and a roof composed of cold rolled copper sheets that are weathering naturally (Figure 3.5).

Blending of Alloy Colors and Finishes

The colors of the various alloys of copper have a natural elegance that can be used to create tonal effects simply by placing one alloy next to another. If mirror polished or satin finished, when purposely assembled the colors of the various alloys can show subtle yet intriguing differences in appearance and character. Figure 3.6 shows a custom brass kiosk for a theater. The brass was left in its natural color and clear coated with Incralac lacquer. Various alloys were used to create a

FIGURE 3.5 The de Young museum, 1.5-millimeter-thick copper sheets made into Inverted Seam® panels. The sheets are of untreated alloy C11000.
Source: designed by Herzog & de Mueron, uses

FIGURE 3.6 Kiosk at the Midland Theatre, Kansas City, Missouri.

subtle contrast. The fluted columns were formed from C22000 wrought sheet. The columns capitals were cast from C87300, the top dome is C28000, and stamped C26000 is used on the entablature and the decorative scrollwork on top of the dome.

The various colors of the copper alloys offer an esthetic palette for the artist and designer. As discussed in the first two chapters, when zinc is added to copper the color becomes more golden yellow. Each percentage increase pushes the color of the wrought or cast form. The addition of nickel takes the color to a golden silver, and the addition of aluminum takes the color to a yellow golden tone.

Figure 3.7 depicts the basic colors revealed in the natural uncoated metal of various alloys given a diffuse, Angel Hair® texture. As these metals oxidize (particularly those with more transparent oxides), the color differences are maintained. Statuary chemical treatments make the differences in color moot, but those underlying differences remain as the alloy constituents react to form the oxide.

Satin and Mirror Finishes

The mechanical processes performed on copper alloys, either as initial finishes for later patination or as final finishes, involve largely the same techniques as those used on stainless steel and aluminum. To arrive at satin or matte finishes, common methods include controlled scratching (such as

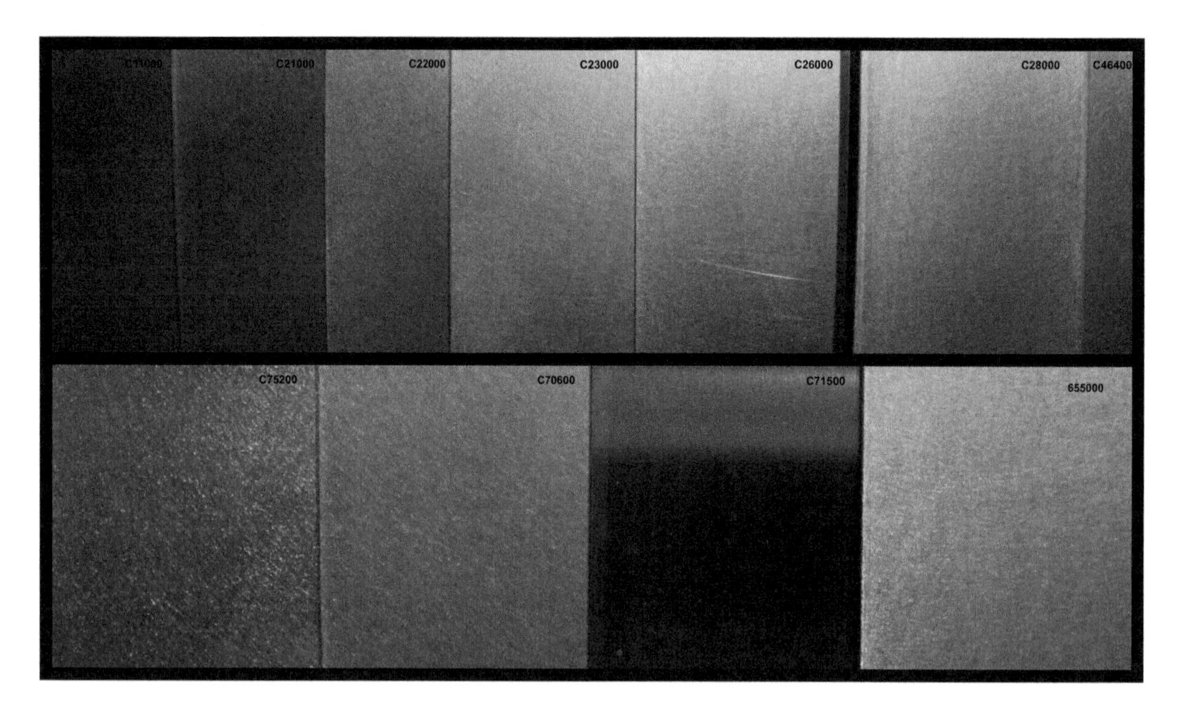

FIGURE 3.7 Comparison of the natural color of various alloys of copper.

FIGURE 3.8 A large cone of the naturally colored C22000 alloy with an Angel Hair finish designed by the RDG Dahlquist Art Studio.

that used in producing the Angel Hair finish), satin directional finishing with Scotch-Brite™ pads, aluminum oxide discs, silicon carbide discs, and glass bead blasting. These all produce levels of diffuse light scattering when light is reflected from the surface. These finishes reveal the real, natural color of the metal. Figure 3.8 shows a large cone made of the welded and polished C22000 alloy. The surface has a fine Angel Hair texture finished with three coats of clear Incralac to preserve the color.

Custom finishing, such as the texture induced onto the C22000 alloy plates shown in Figure 3.9, can produce amazing depth and beauty. The plates were produced using a robot-assisted polishing sander with aluminum oxide discs of 240 grit. The finish is layered and applied to the surface at different angles. The effect can be quite remarkable. The fine grid captures and reflects the light. The surface is coated with a clear Incralac to seal it.

The Scotch-Brite pad is a common finishing component for copper alloys. The pads are color coded from coarse to fine according to their respective comparable aluminum oxide grit: the lower the grit size number (which corresponds to the sieve size), the coarser the polish will be (Table 3.1).

FIGURE 3.9 Curved custom polished surface using the C22000 alloy.

TABLE 3.1 Scotch-Brite pad color and corresponding grit range produced.

Color of 3M Scotch-Brite pad	Comparable grit
Tan	120–150
Dark gray	180–220
Brown	280–320
Maroon	320–400
Green	600
Light gray	600–800
White	1200–1500

Various sanding discs and belts of similar grits can be used in place of these pads to apply directional finishes. Copper and stainless steel wool can also produce light scratches on the surface, although the process is less easily controlled than one using pads, discs, and belts because it is done by hand.

For a mirror finish, polishing and buffing operations involve a step-down in rouges. Copper alloys—with the exception of aluminum brasses, which can be polished, but whose hardness makes abrading the surface more difficult—take well to mechanical polishing because of the relative softness of the surface. (Moreover, the polish on aluminum bronze will be cloudy.) In copper and copper alloys, the polishing and buffing process can develop bright, mirror-like reflective surfaces. Figure 3.10 shows polished copper sheet: the reflectivity is very specular; however, the in-and-out bumps on the surface mute the reflectivity, similarly to how frit alters the light passing through glass.

The mirror polishing of sheets, plates, bars, and tubes is often done in a secondary polishing facility. These facilities use large vacuum tables and have CNC-controlled reciprocating buffers to provide a near-mirror reflectivity to copper and copper alloys.

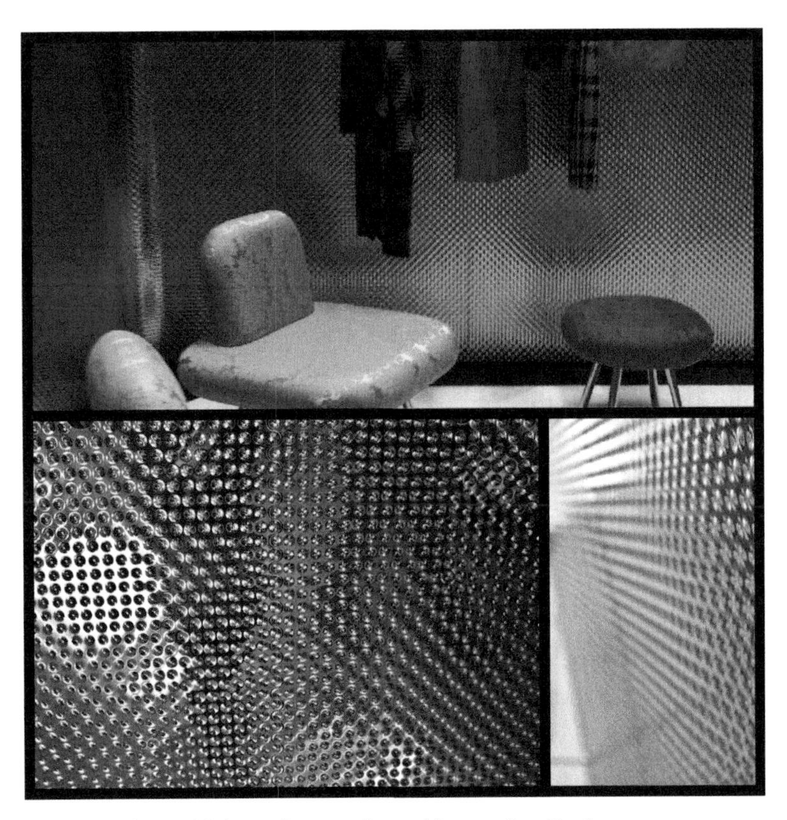

FIGURE 3.10 A C11000 surface with in-and-out embossed bumps in a Prada store.
Source: The designer is Herzog & de Mueron.

Polishing and Buffing

Polishing and buffing operations displace a portion of the moisture at the surface and apply compounds of various oxides of aluminum, iron, and silicon that take the surface down in stages. Polishing is not the same as buffing. Polishing removes metal and is primarily a smoothing and cleaning process that follows abrasion processes. Power equipment is typically used to finely abrade the metal surface and remove small, minute particles on the surface. Polishing belts, discs, and pads impregnated with silicon carbide or aluminum oxide are used with varying degrees of coarseness.

Buffing is a surface refinement process that can be used to produce a particular luster on a metal surface. There are four levels of buffing that take the surface of the metal to increasing levels of refinement, with each level imparting more luster than the previous one:

A. Satin finishing, which produces fine directional or multidirectional lines
B. Cutdown buffing, which produces an initial smoothness on the surface
C. Cut-and-color buffing, which produces a fine intermediate luster on the surface
D. Luster buffing, which produces a mirror-like surface with a high luster

Buffing uses greaseless compounds that are impregnated with various hard oxides of differing hardness. Some of the stainless steel compounds include tallow to speed up the process. The compounds are in liquid or stick form and are applied to various types of buffing wheels made of sewn or loose muslin. After the compound is applied, the buffing wheels rotate at high speed to polish the metal surface.

Buffing compounds are composed of fine abrasives such as tripoli, a microcrystalline silica. Other compounds are called "rouges." Jewelers rouge, also called red rouge, is a compound that is embedded with red iron oxide to give it good polishing action. Green rouge is embedded with chromium green oxide. It is important to test each of these rouge compounds to determine their effectiveness, since there is no standard in their manufacture. There are variations in the size and amount of compound embedded in the bar.

Table 3.2 lists the names and approximate grits of some of the more common cutting and polishing compounds. Tripoli and red rouge are commonly used for copper alloys. Various levels of tripoli

TABLE 3.2 Buffing compounds used on copper alloys.

Compound	Color	Description	Oxide	Grit	Mesh size
Black emery	Black	Removes scratches; rough, fast cutting	Al_3O_2	Coarse	8–24
Bobbing compound	Yellow	Fast cutting for coarse, rough surfaces	$Fe(OH)_3$ SiO_2	Coarse	8–24
Tripoli	Brown	Microcrystalline silica	SiO_2	ˇ	30–60
White rouge	White	Calcinated alumina	Al_3O_2		
Green rouge	Green	Green chromium oxide	Cr_2O_3	Fine	70–180
Red rouge	Red	Red iron oxide	Fe_2O_3	ˇ	
Special polishes				Very fine	220–1200

compounds are used to cut the surface of the alloy and put the initial finish in place. The red rouge will then take the copper alloy surface to a very fine polish.

COLOR FROM OXIDATION AND CHEMICAL REACTIONS

Of all the metals, copper and copper alloys offer the most intriguing, natural, and beautiful color possibilities. The copper salts all possess remarkable and intense colors that we refer to as patinas. The methods by which these patinas develop and the chemical compounds they form are extensive.

Patination of copper is as old as the metal itself. Simply heating the surface of natural copper can produce a rainbow of interference colors, from golds to deep blues. These colors are induced as a thin cuprous oxide forms on the surface when oxygen combines with the copper (Figure 3.11).

Copper and copper alloys have the ability to react with natural substances in the environment. The reactions that take place on the copper surface are really a form of degradation or corrosion. The chemical compounds formed during reactions with these natural environmental substances—or other chemical substances in the case of prepatination—become part of the copper alloy surface. Patinas develop from reactions with the copper surface and form colorful oxides, hydroxides, carbonates, sulfides, sulfates, nitrates, and chlorides. The longer the time of exposure to the atmosphere, the greater the bonds between the base metal and the oxide that develops.

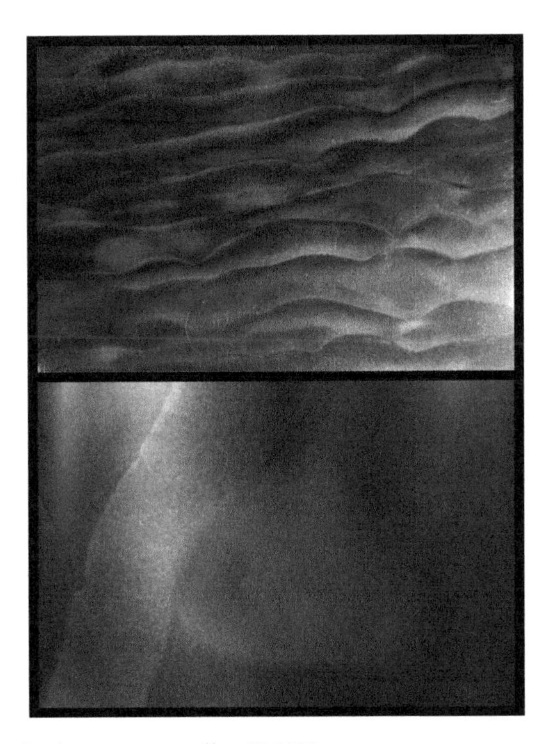

FIGURE 3.11 Heat-generated color on copper alloy C11000.

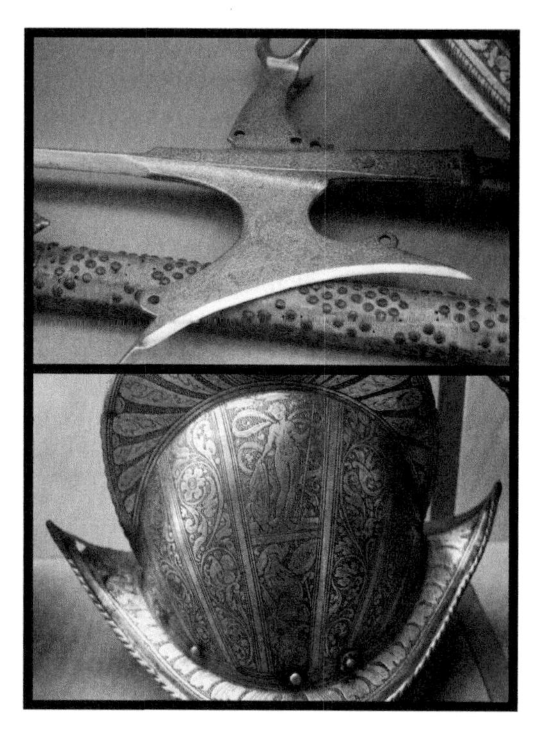

FIGURE 3.12 Niello used on a bronze axe head and a helmet.

Our early ancestors experimented with creating patinas on copper, and later bronze, surfaces. One of the most noted was the black patina known as "niello." This was a black mixture of various substances, including copper, silver, and sometimes lead, mixed with sulfur to produce a paste. This was rubbed into decorative engraved bronze or silver surfaces and then rubbed off, leaving the black substance locked into the grooves. Sometime the niello would reaction with the copper and create permanent contrasting color tones (Figure 3.12).

This blackening of bronze surfaces was used to decorate knife handles, mirrors, and other artifacts, and used to show one's status in ancient Roman and Egyptian cultures as well as in Chinese and Japanese cultures. The black color was induced onto bronze and copper to form the mineral cuprite (CuO), which can be a deep black color. To produce this color the metalworkers of the time had to have an understanding of how to control the corrosion products that form on the surface of copper. Undoubtedly there were other colors these early artisans could induce onto the copper surfaces artificially.

Early uses of the metal by the Sumerians and Egyptians centered around the beautiful colors the copper, and later the bronze, surfaces possess. They could be polished and decorated with engraving, similarly to wood and pottery. These artisans would darken the piece by heating it and then scratch bright copper lines that stood out into the surface. These would eventually darken or fill with dirt and eliminate the contrast in color. The process using niello was preferred because the copper or bronze surface could be polished and the darkened chemicals rubbed into the surface.

Statuary Finishes

Darkening methods that are in extensive use today involve oxidizing the surface to create statuary finishes. Statuary finishes are common chemical finishes used on copper and copper alloys, particularly brass alloys. The statuary finish is characterized by a darkening of the copper alloy surface with linear directional grain induced by small scratches from polishing discs or pads.

Often the term, statuary bronze, is used to describe this finish. There are no precise standards in producing this darkened appearance. The name statuary bronze is used to define a general process.

These finishes are semitransparent, and when properly applied they give rise to a darkened, metallic beauty. Statuary finishes slow down further oxidation by acting as intermediate oxides. Statuary finishes will partially mask the effects of further oxidation on the surface. These finishes are a dark oxide layer produced by a chemical reaction to various chemistries that use sulfides to interact with the copper on the surface. Often the statuary finish is brushed to highlight the finish and remove some of the darkened layer to allow more of the underlying metal to show. Once the finish is achieved, the surface is coated with a clear lacquer or wax. The statuary oxidation process is applied over satin and polished surfaces.

The base metal is first cleaned and completely degreased. A directional or non-directional mechanical finish is applied by hand or with belts or discs. This takes a thin layer of metal off of the surface, including any oxide that has formed and antitarnish substances applied at the Mill source. This is an important step as it places small scratches in the surface that become sites where the chemical reactions can take hold. The surface is darkened by application of an oxidizing compound that usually involves sulfides.

There are several commercial darkening solutions that work rapidly to darken copper alloys. The strength of these solutions can be diluted to slow the process. The darkening is induced by a chemical reaction that forms a dark copper sulfide on the surface. The darkened surface is a transition of the metal into an adherent black or dark brown compound that is chemically bonded to the underlying copper alloy. Typically this is a cold process performed on the copper alloy's surface at room temperature. The chemical darkening can be applied by dipping the copper alloy, or by spraying or wiping the solution onto the surface.

The process involves taking the dark color slightly darker than the final result is intended to be, after which the chemical reaction is stopped by rinsing the solution in clean water. Next the surface is dried, then highlighted by lightly sanding either in a linear direction or in an arching, non-directional texture. The surface is carefully sanded until the desired level of statuary finish is achieved. This is a very subjective process and finishes are usually described as light statuary, medium statuary and dark statuary. These qualities are defined through samples produced on similar metal.

When working with these compounds it is very important to follow environmental and safety procedures. Wear gloves, eye protection, and respirator protection. The chemicals often contain mild acids or strong bases. The dust is very irritating to the eyes and mucous membranes. Work with companies that have established environmental procedures and follow safety protocols.

A statuary finish can be applied to both wrought and cast work. Figure 3.13 shows the darkened surface of the C22000 alloy. It was necessary to arrive at a color tone that provided sufficient contrast

FIGURE 3.13 Ohio Holocaust and Liberators Memorial, Columbus, Ohio, Source: designed by Daniel Libeskind.

to the lettering on the sculpture. Working with the Libeskind design team, a very light statuary finish was created to achieve the desired appearance. The light statuary borders on an interference layer, as light reflects off the base metal and gives a reddish tonal effect that changes slightly depending on angle of view.

Statuary finishes are applied to both copper and copper alloys in all wrought and cast forms. Copper, brass, and bronze alloys are most receptive. It is more difficult to produce the statuary finish in aluminum–bronze, nickel–silver, and copper–nickel alloys because of their corrosion resistance.

It is important to understand that the statuary finish is achieved through a chemical reaction with the metal surface; it is not a coating in the sense of a paint or lacquer. Components in the particular alloy will cause a slightly different color tone as the oxide develops from reacting with the elements on the surface. In order to convey information about the potential color, create samples using exactly the same alloy, which will allow you to arrive at a target color tone to work toward.

Figure 3.14 shows large brass doors made of different alloys. Each alloy was treated in the same sulfide bath, but due to the variation in translucency among the alloys the colors are different. The result is remarkably beautiful, with subtle differences developing on the surface.

A statuary finish can be applied to large surfaces to darken them, usually using a dipping process. The length of the time of immersion in the darkening solution is critical because highlighting large quantities to arrive at a lighter tone is not easily achieved.

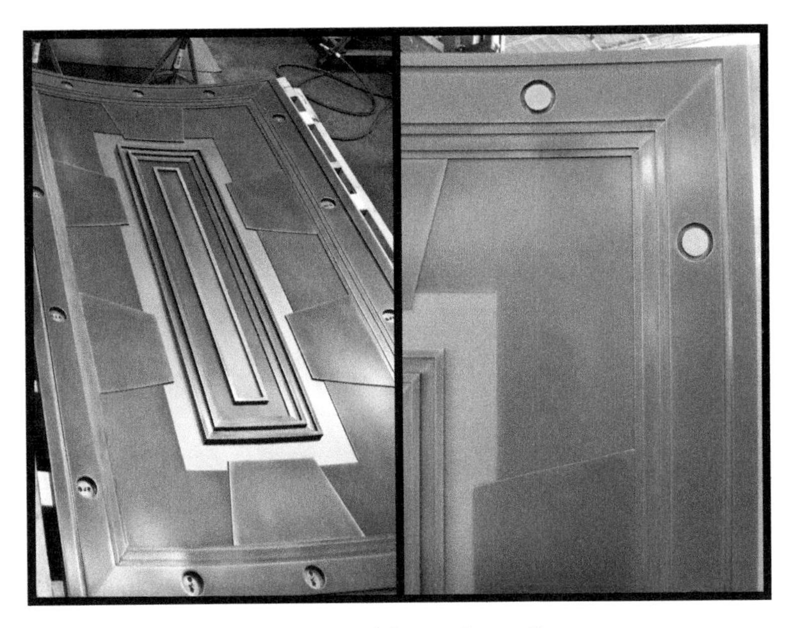

FIGURE 3.14 Custom doors made of the C22000 and C28000 brass alloys.

What typically occurs is the darkening solution is diluted to allow reaction time. Temperature as well as strength of solution is monitored closely because each subsequent dipping reduces the effectiveness of the remaining chemistry in the bath. It must be understood that the process is a chemical reaction, and that chemical reactions by their nature involve numerous variables in the effort to achieve any semblance of repeatability.

Critical variables in the process of applying statuary finishes are:

- Alloy
- Initial surface condition of the metal
- Concentration in solution
- pH of the solution
- Temperature of solution
- Quality of water
- Quality of rinse tank water
- Time of immersion
- Sulfide solution used
- Method of final highlighting of the finish

Once a target finish is established, the darkening of a series of elements to make a surface is highly dependent on controlling the above variables. If the variables are established in the target

FIGURE 3.15 Bond Street window surrounds made of the C11000 alloy,
Source: designed by Herzog & de Mueron.

finish, the two keys are concentration of the solution and of the rinse water solution, which change throughout the process as metal is dipped into the tanks. Operator skill is important: the operator needs to recognize when it starts taking more immersion time to achieve the same results, because this is an indication that the solution concentration has changed. This issue is the same for the rinse tank, where the solution concentration can increase and continue the darkening.

Figure 3.15 shows the copper alloy C11000, which was darkened and formed into a bell-shaped cross-section. The dark tone was created using an ammonium sulfide solution. Ammonium sulfide will rapidly darken copper alloys, so it needs to be diluted to control the process. When working with ammonium sulfide, exercise extreme care as this powerful base has a pungent odor and should be disposed of properly and carefully. Another common darkening solution is made from potassium sulfide, also known as "liver of sulfur." This solution is a favorite of patenuers as a darkening base for many other patinas. It works well when applied hot and when applied slowly, building up the dark tones. Solutions made from potassium sulfide have short shelf lives and need to be stored in opaque containers due to their light sensitivity.

A darkened statuary finish can give copper alloys a deep, old bronze appearance (or at least what we today associate with a bronze finish). The term "statuary bronze" is an idiomatic term associated with this type of finish, but the subjective nature of how this finish is used, and how the variables are adjusted is more art than science. The left-hand image in Figure 3.16 shows C22000 plates with a statuary finish produced using a commercial copper darkening solution containing an oxidizing

FIGURE 3.16 *Left*: Addition to the Walker Art Building at Bowdoin College in Brunswick, Maine. Designed by Machado Silvetti. *Right*: A sculpture at the Farmland Industries building made of the C11000 alloy. Sculpture by L. William Zahner.

agent and a sulfide compound. The image on the right shows C11000 with a statuary finish created by a potassium sulfide solution.

Matching Hardware Finishes

To add further confusion, in the hardware industry descriptions for various statuary finishes such as "oil-rubbed bronze" and "antiqued brass" are common. These are terms given to describe a range of possible finishes created by various manufacturers of hardware components such as door knobs, locksets, push plates, faucets, hinges, and similar stamped or machined products. One company's oil-rubbed bronze can be considerably different from another's, and the same can be said about antiqued brass and antiqued bronze finishes. These finish names are derived from the hardware and fixture industry, which uses them for copper alloy or copper plate surfaces treated with a darkened rich tone induced into or applied to the surface. Although the name of the finish may be the same, the color is unlikely to be the same across producers, and attempts to match a wrought plate or sheet material with a cast hardware or fixture by specifying an oil-rubbed bronze finish is destined to fail, as the variables are many. As with all oxides and patinas, these finishes are not paints but chemically reacted surfaces of the specific copper alloy makeup. The first variable lies in matching up as closely as possible the alloys used. If the hardware is created from machined or cast parts, the alloy will diverge from what is available in wrought sheet or plate, as the alloys will have been produced for different processes and different industries.

As mentioned above, terms such as “oil-rubbed bronze” and “antiqued brass” are idiomatic. They represent one manufacturer’s particular finish, not a standard finish. The same is true of light, medium, and dark statuary bronze finishes. These finishes depend on the alloy used, how the particular chemistry reacts with the surface of the alloy, and the technique of the particular factory producing the finish.

The oxidation process used to darken these oil-rubbed and antiqued finishes is a specific one that also needs to be matched closely in preparation, application, and temperature. Most hardware companies have proprietary systems and processes for putting these finishes on their products. The processes may not be transferable to wrought forms of copper alloys. For instance, there is no actual oil used in the oil-rubbed finish. The term harkens back to a time when linseed oil or paraffin oil was used to protect a surface. There was a time when a form of mineral oil, such as paraffin oil, was rubbed into an oxidized brass surface to produce a sheen and slow the oxidation process. No such process is used today. The term is a holdover from the past and generally refers to a highlighted statuary finish that has been lacquer sealed.

To achieve a match for these hardware finishes, you must use a process of trial and error to arrive at a color tone close to the hardware. This can be a tedious and difficult process. The copper alloy part is thoroughly cleaned of all grease and oil, and oxides are either removed mechanically or dissolved in acidic solution. The part or surface is darkened to a deep black or charcoal tone and then brought back by light sanding to bring out highlights in the surface. The oil-rubbed finish is typically darker and matte gloss, while the antiqued surface has typically been highlighted and is usually the more reflective of the two. No two manufacturers have the same oil-rubbed or antiqued finish, so the work of matching parts such as handrails and elevator panels by matching the actual hardware must be performed. Doing the reverse is not possible, because a particular hardware manufacturer will have set alloys and processes that can’t easily be adjusted for other wrought or cast forms.

In the context of building hardware, an oil-rubbed finish is expected to age naturally and is not intended to be coated or protected. In the past the building hardware industry used a finish number of US10B for oil-rubbed bronze; later this was changed to BHMA613 by the Building Hardware Manufacturing Association, or BHMA. The BHMA has a set of representative samples to enable you to develop a finish within a general range of what the real finish is, as supplied on locksets, door handles, and other hardware used in building construction. The American National Standards Institute (ANSI) has partnered with BHMA to provide a method for specifying these finishes. ANSI/BHMA Standard A156.18-2000, *Materials and Finishes* describes US10B/BHMA613 finish as a dark oxidized satin bronze, oil rubbed. This still leaves broad latitude for interpreting what finish this really is.

As discussed in Chapter 2, there are a number of alloys that fall into the brass and bronze family. They all will react differently to oxidation chemistry and develop different colors.

Appendix B has a list of copper alloy hardware finishes with basic descriptions. Presenting an actual metal sample with the specification is highly recommended. Hardware producers have set methodology and processes that are difficult to adjust; a custom metal manufacturer can adjust a process by using different alloys, adjusting the chemistry, and highlighting the surface to achieve a close match. It is rare that an exact match can be achieved, so the expectation should be to achieve a finish that appears to be in the same family and is an approximate match.

When specifying the statuary finish, consider the following:

- Provide an image of the finish. If possible, have a physical sample available to view.
- Describe the finish along with providing the pertinent hardware code or other identifying number. This establishes the finish as a chemical process (some less sophisticated metal manufacturers might substitute a faux paint for the chemical process).
- Include a description of the alloy. If you're trying to just match a finish, then leave it to the manufacturer to provide the appropriate alloy to achieve the color.
- Require light, medium, and dark samples to establish a target range for color.
- If the color is somewhat off, it may be the alloy used. The manufacture should be able to access other alloys to achieve a better match.
- If the tone is darker or lighter than the target color it can be adjusted by modifying the level of highlighting. The manufacturer should be able to adjust this.

Dirty Penny

There are several color variations available in the statuary finish category beyond those generated out of the building hardware world. One that has found use on a number of architectural surfaces is the Dirty Penny finish™.[1] This finish is produced on copper or on alloys of copper with a high copper content. The colors are mottled and initially have a slight iridescence. Figure 3.17 shows various tones of this finish as developed on the C11000 alloy.

The Dirty Penny coating is a transition coating analogous to the statuary finishes but more mottled and indistinct. It is very durable and can be used on both exterior and interior surfaces. The level of iridescence is variable and will eventually darken out in exterior projects, as some of the iridescence disappears as the oxide thickens. Figure 3.18 shows a Dirty Penny finish in New York City after being exposed for three years. The mottled contrast is still there, but the iridescence has been muted as the surface absorbs moisture from the air and the natural cuprous oxide thickens over the entire surface.

Creating this patina is similar to creating the red patina on copper. It is a very durable patina and has been used on tabletops, ceilings, and wall panels. In Figure 3.19, the top image shows a perforated and bumped wall used in a garden in Korea. The iridescence of the finish is still apparent

[1]The name and the process for the Dirty Penny finish came out of a meeting with the famous designer Antoine Predock. He wanted a finish for large surfaces that resembled an oxidized penny (pre-1984, when pennies were made of copper versus the plated pennies of today). From this meeting the Dirty Penny finish was "coined."

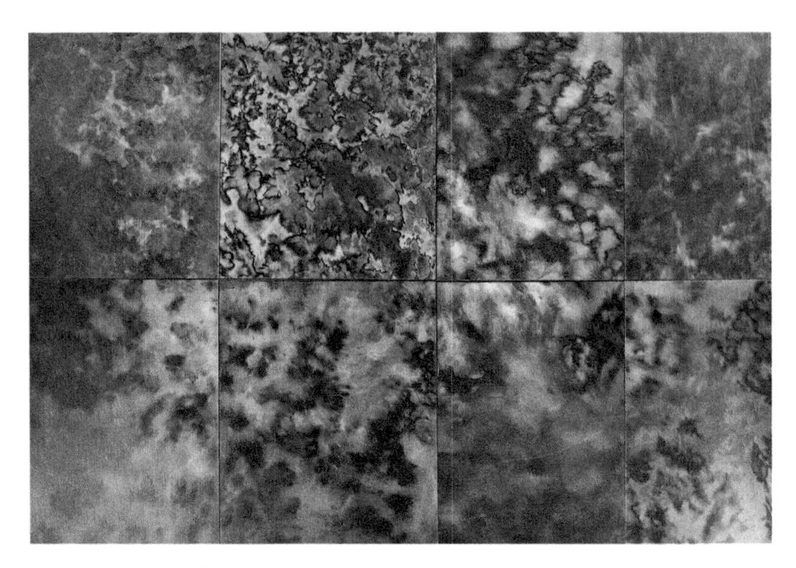

FIGURE 3.17 Various Dirty Penny tones.

FIGURE 3.18 Dirty Penny on copper panels made of the alloy C11000 after three years of exposure in New York City.

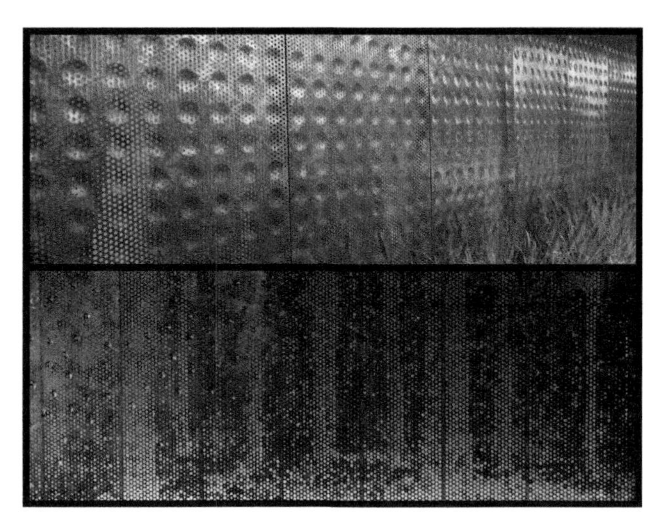

FIGURE 3.19 Perforated and bumped Dirty Penny copper alloy C11000.

after two years of exposure to Korea's environment and climate. The bottom image shows a perforated and bumped surface used on an interior signage wall, with contrasting lighting that shows a "forest of trees."

Black Patinas

Black patinas have been used on copper alloys from the early days, allowing artisans to highlight and provide contrast in bronze and copper fabrications. These artisans would either etch into the blackened plate so that the metal could show through or reverse the process, leaving the etched portion black while polishing the balance of the artwork.

Alloys of commercially pure copper, brass, and bronze (with the exception of aluminum bronzes and phosphor bronzes) can be blackened. The blackening process can be an extension of the statuary process. The blackening is carried out on the surface to an extreme level, which produces a deep, rich color. The artist John Labja took a plate of commercially pure copper 50 mm thick and waterjet cut it to exactly 1829 mm square (72 in.), then treated it to remove the scale and surface oxide. The clean copper plate was exposed to a potassium sulfide and copper sulfide solution for several minutes, then given a polish and rinse. The color achieved was a deep black on all surfaces (Figure 3.20).

Depending on how the process is applied, you can eventually get a chalky white substance developing over the black in some areas as copper carbonate develops and the slightly lighter surface shows as a contrast to the deep black.

Figure 3.21 shows copper alloy C11000 plates that are 2 mm thick used on the Robert Hoag Rawlings Public Library in Pueblo, Colorado. After 16 years of exposure, there are indications of a slight whitening in different places on the black patina surface. The surface was oxidized using a potassium sulfide solution along with a strong oxidizing agent.

FIGURE 3.20 Fifty-millimeter thick black copper plate by John Labja.

FIGURE 3.21 Robert Hoag Rawlings Public Library, Antoine Predock Architect.

Other methods of producing a deep, durable black color involve immersion of the copper sheet in a hot bath of sodium hydroxide. This creates a thick layer of cupric oxide on the surface. At a microscopic level, tiny crystals set perpendicular to the surface and give it a matte appearance that catches and traps light. A very durable black texture can be created using this method.

In Figure 3.22, the top image shows the American Heritage Center and Art Museum designed by Antoine Predock and clad in blackened C11000 sheet. The lower left image shows a roof on a residence designed by the architectural firm BNIM; the roof is clad with blackened C11000 alloy. The image at the lower right shows a handrail end at the Smithsonian's National Museum of the American Indian, also made from C11000 alloy bar and welded to a machined plate.

This durable black finish absorbs solar radiation and can get quite warm. Used on a roof surface, the roof will shed snow and ice more rapidly as it will heat up quickly in the sun. When a surface with this finish is first installed, the crystals will be long. Running your hand over them will knock them down and can change the appearance. If the metal is placed where it can be touched or if handling is extensive, you may want to coat the surface or rub it to make it match. The handrail in the lower right of Figure 3.22 was coated for this reason.

FIGURE 3.22 Blackened copper alloys in three projects.

Red Patinas

Red is a very difficult color to produce in copper alloys. Red is the color of cuprous oxide, so one would think this would be a common, achievable color in copper alloys. However, it is not a simple patina that is always stable. Figure 3.23 shows various red patinas. The image at the upper right shows a vase created from hammering high-arsenic copper. As the metal is repeatedly heated to anneal and soften it, and then hammered to create the shape, the color obtained can be a deep red. The object in the upper-left image is a deep red that was created by freezing the cuprous oxide that forms when heated under molten glass, which locks the copper surface into a frozen, airtight form. The bottom image shows the C65500 alloy treated with a hot copper sulfide solution until the metal turns red. The color has questionable stability unless the metal is coated to seal it from the environment.

The red finish, when achieved, is rich and beautiful. As discussed, it will change over time and darken if not protected with a clear coating. Figure 3.24 shows a beautiful example from Korea, with 2-millimeter-thick copper plates oxidized to a red tone and then allowed to age. The copper was not

FIGURE 3.23 Three examples of red patinas.

FIGURE 3.24 Red patina on panels at the Daeyang Gallery and House, Korea, Source: by Steven Holl Architects.

coated, and the red has darkened in some places and developed interference colors of purple on portions of the surface after a 10-year exposure.

Green Patinas

The term "patina," as it pertains to copper alloys, originates in Europe. In Italian, the term *patina* means a thin layer of deposit on the surface of an object. It is derived from the Latin term *paten*, meaning a shallow dish or pan. Today the term has taken on a more expansive definition and means any desirable color tone on the surface of any metal. The deep oxide on weathering steel and the oxides on zinc are both referred to as patinas. In this context, a patina is different from tarnish or rust, each of which would be construed as an undesirable surface condition.

A corrosion engineer or a metal conservationist might define a "patina" as a destructive layer of oxide that forms on a copper alloy surface through thermodynamic interaction with the atmosphere. The rich green color on a copper roof and the oxidation of a bronze sculpture in a fountain are both manifestations of the same, natural progression of surface decay. Although the results are similar, however, the consequences are far different. The copper roof is intended to oxidize and grow a beautiful green tone that resists further oxidation, as shown in the roof of Børsen or the 1909 spire on the Nikolaj Contemporary Art Center (Figure 3.25). The sculpture, on the other hand, is undergoing surface changes that are irreversible and can be damaging to the original design intention.

FIGURE 3.25 The patinated copper roof dating to 1625 on Børsen and the 1909 spire on the Nikolaj Contemporary Art Center in Copenhagen, Denmark.

Patinas are a modern human-induced change to the copper alloy surface. Early sculpture produced by the Greeks and Romans were not patinated but instead were often polished and sometimes coated with pigments or with gold. The Romans used the blackening and highlighting process involving niello mentioned earlier in the chapter. As described, the artist would engrave the surface and fill the artistic design with niello, a black mixture of copper, sulfur, and lead. The surface would be burnished and polished, leaving the contrasting design in the polished bronze surface. The bronze sculptures of antiquity were often polished and are believed to have been selectively colored (for example, by coloring the lips or eyes to add contrast) and enhanced with other materials afterwards.

The industrial age that introduced combustion products, first to heat our homes and factories and next to transport us about, sent an abundance of sulfur into the air. Copper surfaces interact with sulfur and capture it as a copper sulfate.

There certainly were many bronze sculptures set near the sea in ancient times and these would have developed the characteristic green of hydrated copper chloride. The sculptures are thought to have been coated with pine tar or bitumen to not only produce an overall tone but also to protect them from chlorides from the sea.[2]

Figure 3.25 shows a 400-year-old copper roof in Copenhagen, Denmark that is still performing both esthetically and functionally. The spire shown in the image on the lower right was built in 1909 and now adorns the Nikolaj Contemporary Art Center. Patinas of rich mineral forms of copper such as the one on the spire were not induced; over time the copper pulled the sulfur from the air and created a beautiful, natural surface. This type of patina forms a very strong bond between the oxide and the base metal. The crystals that have formed on the surface are large and intertwined with the copper base metal. The green patina we see on the copper roofs of decades-old buildings is composed of the mineral brochantite, which has the chemical formula $Cu_4SO_4{\bullet}(OH)_6$. This is a hydrated salt of copper composed of sulfur. In the past it took years to develop but today, with our less polluted atmosphere, it may take decades, if it forms at all. Sometimes there are carbonates and chlorides intermixed in these natural patinas. Most likely this is the case of the Børsen roof in Copenhagen (top and lower left images). Where surfaces are installed near the humid seacoast, copper chloride salts will be the predominate mineral forms. There are some that would argue that the copper of today has a more refined grain structure than the copper of old and that the older metal was more inhomogeneous in its makeup.[3]

Natural copper patinas alter the thermodynamic behavior of the copper surface by altering the chemical composition on the surface, making the surface more inert by coating the copper with a nonreactive mineral substance. These patinas are very adherent and are chemically bonded to the copper itself. If one were to remove portions of the copper sheet and subject it to mechanical forces such as bending, the patina would crack. It does not possess the flexibility of the base metal.

[2]P. Weil, "Patina from the Historical-Artistic Point of View," Northern Light Studio website, accessed October 17, 2019, http://www.northernlightstudio.com/new/patina.php.

[3]C. Leygraf et al., "The Origin and Evolution of Copper Patina Colour," *Corrosion Science* Vol. 157, pages 337–346 (2019).

Natural patinas are formed by the interaction of moisture, pollution, chlorides, and carbon dioxide on the surface of copper and copper alloys. They form over time and with exposure and benefit from humidity in the ambient conditions. Moisture is needed to develop these patinas through developing electrolytes on the surface. Add in pollution and the copper develops bonds with sulfur, chlorides, and carbonates. These patinas are essentially natural filters of airborne pollution, as they capture and lock these substances into mineral compounds.

Artificial Green Patinas

Artificial patinas are a decorative treatment to copper alloy surfaces. These patinas are planned chemical reactions on the surface between the copper and chemical compound. The surface forms integral chemical bonds between the copper and alloying elements with oxidizing compounds. Artificial patinas mimic what occurs in natural exposures, but they are produced in a concentrated and controlled environment in order to arrive at a predictable outcome and a specific color tone, and to dramatically accelerate the time of development. The main difference is the crystals that make up the patina. These crystals have formed quickly and are smaller than what forms naturally over long periods of exposure.

Today the prepatinated copper sheet is a common product available for roofs and walls (Figure 3.26). With the quality of air today, copper will take a significantly longer time to develop the characteristic green patina of old. Deep green natural patinas, which used to take a decade or so to develop, now will take close to a century or more. Because of this designers have sought to accelerate the process by utilizing prepatinated material to get the color tones from the beginning. The process does not use paints or dyes but the actual chemical process that nature would over years of exposure.

These patinas can perform well in all exposures, but their preparation and execution require expertise. Moisture and oxygen, along with carbon dioxide, will enrich these surfaces very slowly over time and exposure, making the patinas similar to naturally forming patinas. When first installed, the patina will exhibit the color tone produced in the factory. After a few months of exposure, the patina should develop a richer tone as it absorbs moisture from the air as well as from carbon dioxide and other pollutants. Figure 3.27 shows prepatinated copper walls when first installed in the upper image and as they have weathered over a decade in the lower images.

Roof surfaces are exposed to rains, hailstorms, ultraviolet radiation, and particles of pollutants coming out of the atmosphere. These tend to remain on the surface for a prolonged time period. As moisture from condensation forms, these pollutants can combine with the copper patina and darken it. Figure 3.28 shows an urban roof in Hong Kong made of prepatinated copper with a soffit overhang. Both have held up well. The roof has been in place in a humid urban environment for over 20 years.

There are numerous formulas for creating artificial patinas on copper and the various copper alloys and there are several proprietary mixtures available to induce rich patinas on copper alloy surfaces. Successful patination of copper surfaces can be particularly capricious due to slight variations in the metal surface, the temperature during and after patination, humidity, the strength of

FIGURE 3.26 Walls clad in prepatinated copper.

FIGURE 3.27 Patinated walls right after installation and several years later.

the chemistry, and an extensive number of other conditions that must be planned for and adjusted for. Controlling these variables in order to achieve a measure of success is extremely difficult, particularly for large architectural surfaces. It is a process that cannot be rushed. It is not paint but a chemical reaction of the base metal with an oxidizing agent and either sulfur or chloride. Figure 3.29 shows several varieties of patinas applied to copper alloys.

Artistic expression can be enhanced as well by the deep, rich patinas created on smaller surfaces. Figure 3.30 shows various artworks in which smaller elements were patinated in a process similar to that used on large elements that form a surface. Here the artwork was carefully prepared and the patination solutions applied slowly and in a controlled environment until the desired color was achieved. In each case a transition coating of dark oxide is often first created before the green patination solution is applied and allowed to react.

FIGURE 3.28 Roof and soffit surfaces in Hong Kong.

Often when the green patination solution is first applied, the uninitiated find quick success as the surface changes and the green tones begin to develop. As with paint, the impulse may be to add more to thicken the coating, but if this is done the first reaction will be prevented from taking hold and will be choked of air as the additional solution reacts with the copper ions on the patina surface, and the patina will lift from the base metal. Once dry the fragile patina will flake off. Since there are no dyes or pigments in patination, a chemical reaction with the copper is required (Figure 3.31).

A chemical reaction takes time: time to allow the copper ions to diffuse out and combine with the oxygen to form an initial layer of cuprous oxide. As the reaction continues, the copper ions join with sulfides, chlorides, and other substances entering the moist layer and becoming ions in solution. Prepatina work in a plant usually involves an oxidizer that strips the copper of electrons and speeds up the diffusion into the solution containing sulfide or chloride ions.

FIGURE 3.29 Various patinas that can enhance copper alloys.

The environment needs to be one of tight humidity and temperature control, which is one reason why in situ patination of large architectural surfaces is extremely difficult and prone to failure. Additionally, the chemicals used are not ones that you simply rinse down the drain; they must be contained and dealt with properly. In situ patination can damage the local environment. Factory patination should be performed only by those that have proper disposal systems in place. Otherwise you are sacrificing the environment to achieve an esthetic. It is not worth it.

Artificial patination of copper alloys is essentially forced and controlled corrosion. You are speeding up the natural process that would normally require years to develop. It is not a paint coating, but a surface induced by chemical reaction. You can no more precisely control the patina development on a copper alloy surface than you can predict the exact formation of bark on a tree.

FIGURE 3.30 Various artworks made of patinated copper.

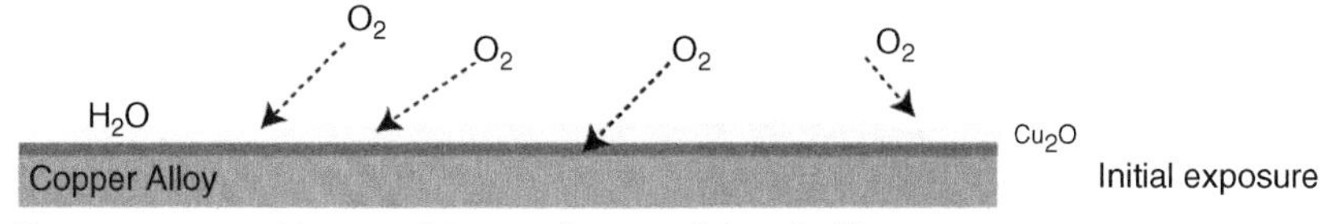

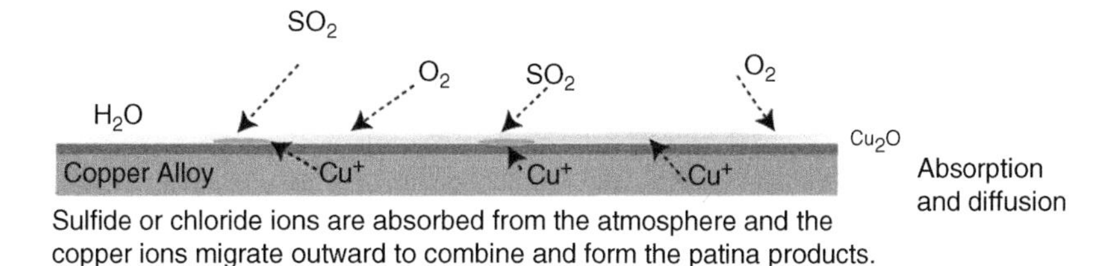

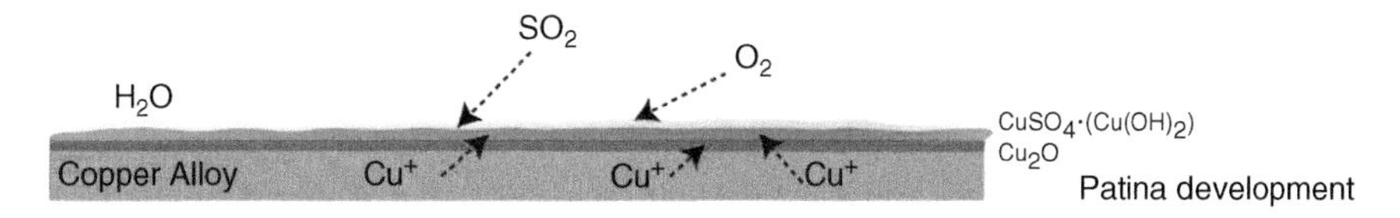

FIGURE 3.31 Formation of patina.

Additional Patina Colors

Colors produced by patinating copper alloys are dependent on the chemicals used to cause the reaction and the makeup of the metal at the surface. You can use the same chemistry, temperature, and humidity on two different alloys and the results will be different. The color tones will be different from one alloy to the next.

The process of producing a patina is an art. The techniques are often trade secrets or proprietary processes gained through much experimentation and trial. There are as many failures as there are successes in patination processes on copper alloys. Most patanuers, once they have perfected a given system or process, tend to stick to it. The variables that must be controlled are many, and larger surfaces, architectural walls, and roof surfaces can pose particular challenges due to the need for levels of consistency in the finished product and the time necessary for production.

The patina development can be layered. One intriguing effect results from combining various chemistries and applying the patina solutions in different waves. Some solutions react quickly with the copper alloy surface while others may take days. Figure 3.32 shows the layering of patinas in the copper panels used for the Smithsonian's National Museum of the American Indian.

FIGURE 3.32 Layering of patinas for the art wall in the Smithsonian's National Museum of the American Indian.

TABLE 3.3 Colors achieved using certain chemical compounds.

Color	Chemical compound	Difficulty
Black	• Potassium sulfide • Ammonium sulfide • Hot sodium hydroxide	Easy to achieve. Chemicals have pungent odors and short shelf lives. There are a number of proprietary mixtures.
Brown	• Ferric nitrate • Potassium sulfide • Copper sulfide	Easy to use. Various concentrations. There are a number of proprietary mixtures.
Antiqued	Potassium sulfide	Easy to use. Several proprietary mixtures are available.
Green	• Copper acetate • Copper carbonate • Copper nitrate • Copper sulfate	Several proprietary mixtures are available. Hot and cold versions. Can be layered with potassium sulfide or ferric nitrate solutions to adjust color.
Blue green	• Copper nitrate • Copper chloride	Temperature sensitive. Can be combined with ferric nitrate solutions to adjust color.
Yellow to orange	Ferric nitrate	Altering the strength can push the color from golden brown to yellow.
Red	• Ferric nitrate • Copper sulfide	This is a hot application. Can be difficult to achieve.

A few of the colors that can be achieved from various chemical solutions are listed in Table 3.3. These are in common use by the professional patanuer.

When working with patinas, exercise caution; many of these processes involve chemicals that can be a hazard to a person's health and the environment. Follow carefully all health and environmental regulations, wear protective face and eye covering, and only work in well-ventilated spaces. Dispose of all rinse water and excess chemicals correctly.

There are two distinct categories of patination processes, the cold application and the hot application. The cold application process is used most often on wrought forms of copper, such as sheet and plate. A cold patina was used to create the beautiful surfaces on the National Museum of the American Indian art wall (Figure 3.32). This finish can be layered and applied to create a mosaic across a surface. Figure 3.33 shows the creation of two such wall surfaces with the C11000 alloy.

The hot application process is predominantly used on castings and sculpture due to the potential for warping thinner plates and the ability of thick castings to distribute the heat from the process. In both processes there is a consistency in how the surface is prepared. A surface that is clean and free of all oils, soil, and water is critical for a successful patina to develop. Starting with a clean surface can often be one of the more difficult steps in the process of patination.

FIGURE 3.33 Custom patination across multiple panel elements.

Preparing the Surface

It is important that the copper alloy surface be as clean as possible. It must be free of all grease, oil, fingerprints, moisture, and other organic substances. It also needs to be clean and free of all oxides. As described previously, copper alloys react rapidly with their surroundings, and any oxidation on the surface can render the alloy unreactive. For patinas to be effective, the surface must want to react with the chemistry of the new localized environment.

When degreasing the surface, start out simple. Wipe the surface down with 99% isopropyl alcohol and allow it to dry. There are several commercial degreasers that will also work well. Some contain acids that aid in the breakdown of the oxide coating, while others are alkali treatments with surfactants to carry the surface substances away.

For large areas use a solution of trisodium phosphate wiped over the surface. Trisodium phosphate can harm the environment if allowed to get into streams or lakes, and precautions should be followed to avoid skin and eye irritation. Use it sparingly by sponging it on the surface.

Oxides can be removed from copper surfaces by wiping the surface with phosphoric acid. If there are heavy oxides on the surface, they may need to be removed mechanically or by means of stronger acid baths.

It is often desirable to roughen the surface by sanding or light blasting. This removes a portion of the outer layer of the metal, exposing oxide-free metal for reaction and an increase in microscopic surface area by increasing areas where chemical reactions can occur. Roughening the surface can also trap oils, particularly from fingerprints, in the microscopic grooves in the surface.

Once the surface has been degreased, oxides have been removed, and the surface has been roughened it is ready for patination. It is critical, however, not to allow the surface to remain untreated for long. At this point, the surface is very receptive to oxidation and fingerprinting.

Start with a clean surface:

A. Remove all grease, fingerprints, and moisture.
B. Remove surface oxide, tarnish, and old patina.
C. Roughen the surface by sanding or light blasting.
D. Immediately begin the patination process.

The Cold Patination Process

Cold patination is done without a torch or other heating elements. It is often worthwhile to have the metal warm but not excessively so. You need the patina solutions to remain on the surface long enough to react. If it dries too rapidly, a chemical reaction will not occur and the patina can flake from the surface. When this occurs, the process will require reworking, usually on the entire surface.

Building the patina by first producing a transition layer or conversion coating is recommended. This can be a dark sulfide produced by exposure to potassium sulfide (liver of sulfur) or the stronger ammonium sulfide. There are other commercial conversion solutions that can be diluted to adjust their strength; when using these, it is important to allow the intermediate coating to dry thoroughly before wiping it down with isopropyl alcohol or a deionized water rinse to remove all loose substances before application of the cold patina solutions.

Cold patination is best suited for flat sheets or products and shapes created from wrought forms of copper alloy. The patinas can be rolled, sprayed, or brushed on the surface. Cold patination requires temperatures in the 15–27 °C (60–80 °F) for best results: at colder temperatures, the chemical process is slow to react, and at warmer temperatures the solutions dry out before they can properly react, in which case they will flake from the surface. Having the humidity relatively high, around 50–60%, is recommended as well.

In some cases, the color obtained will be different for different temperatures and different relative humidity, particularly when the formula is a blend of compounds and acids. Temperature and humidity control are essential when applying cold patinas.

Cold patinas can be layered to create different effects. You can blend the patinas onto and into one another. In Figure 3.34, the cold patina was applied over a series of sheets to create a mosaic of color. Sometimes, the final color results do not show for hours, even days. Trying to push the process too fast will lead to failure as the beautiful patina flakes from the surface.

It should be noted as well that patinas are essentially inelastic mineral films. Forming the metal after patination will lead to cracks at the edges as the underlying elastic metal forms and stretches the inelastic patina on the surface. If there is major forming to be performed on the metal sheet or plate, then postpatination will be necessary.

Frank Lloyd Wright used a specific chloride-based patina formula on the Price Tower in Bartlesville, Oklahoma (Figure 3.35). Its beautiful patina was created by applications of an

FIGURE 3.34 Cold patina blend over a surface.

ammonium chloride solution. Two applications were used spaced over a period of time. It was an in situ application and the difficulties with application are visible, with areas missing the patina altogether. This is a difficult formulation to use successfully and the success of the patination had a lot to do with the humidity and metal preparation. Today we can see darkening in areas, where cupric oxide has developed. Still, the beauty is in the variations on the surface patina and the way it has gracefully aged.

Cold patinas can be left to age naturally or coated with a clear lacquer or wax. The lacquer can often assist in binding fragile patinas to the surface (Figure 3.36). The sculptural wall shown in the figure was created from 2-millimeter-thick copper sheets hammered and patinated to create the texture and color. Protected with three layers of Incralac, this surface has held up well for over 20 years. It is best to allow the green patinas used on roofs and walls to age unprotected. They should bond with the copper alloy base metal.

The Hot Patination Process

Hot patination is a process commonly used on sculpture and castings. A heat of approximately 90–100 °C (200–220 °F) is applied to the metal. A torch or heat gun is used to bring the metal up in

FIGURE 3.35 Price Tower, Bartlesville, Oklahoma, designed by Frank Lloyd Wright.
Source: Photo by M. Waits, http://Shutterstock.com.

FIGURE 3.36 *Heartland Harvest*, Kansas City Board of Trade,
Source: designed by Joel Marquardt and Gastinger Walker Harden Architects.

temperature while the patina solution is sprayed or brushed on the surface. As with the cold patina process, the solutions can be layered as the surface is reheated. In the hands of an expert, stippling effects and other artistic techniques can greatly enhance the cast sculpture.

Patinas generated from hot patination processes are generally sound and more durable than those created through a cold patination process. Where cold patinas are opaque, hot patinas are more translucent, allowing the metal to reflect through. The chemical reactions take shape more quickly when heat is introduced. The process requires skill and technique to avoid overheating the surface. Skill is also necessary to blend the patina evenly over large areas: too hot and the patina dries out; too cool and the chemical reaction is slow to proceed.

Hot patination is not well suited for flat sheet material or even plate material. The heat can cause the metal to warp and expand. Hot patination can be used successfully for formed parts if they are not too thin. For castings, however, the mass takes heat well. There are several patinas specific to casting alloys. The older alloys once more commonly used in sculpture, such as Gunmetal (alloy C83600), offered rich opportunity for colors. Silicon–bronze alloys C87300 and C87600 have a number of patinas that work well for interior and exterior applications, and even fountains.

There are hot patinas better suited for fountains. Figure 3.37 shows a green patina that is a variation of a patina called "old world green." This hot patina is created by layering a solution of

FIGURE 3.37 Green patina on *Muse of the Missouri* by Wheeler Williams (1963).

FIGURE 3.38 *Left*: Gold patina cast door Source: by Robert Graham. *Right*: *Blackbird* Source: by Larry Young.

ferric nitrate with a solution composed of cupric nitrate. It is very stable and common for outdoor sculpture, particularly sculptures in fountains.

Nearly all sculpture patinas are applied at the foundry where the metal is cast. Different foundries are skilled in the use of various patinas but will work with the artist to achieve a given color tone and effect. The left-hand image in Figure 3.38 shows stunning cast doors at the Los Angeles Cathedral. These massive golden-hued cast doors were designed by the sculptor Robert Graham. The patina applied most likely had a ferric nitrate mix, which gives a rich golden bronze color. The right-hand image is a stunning form designed by the artist Larry Young. The black patina has a rich luster created by a potassium sulfide solution.

Once the casting is pulled from the mold, welded, ground smooth, and cleaned it is ready for the patina. The patinas used are complex oxides created by carefully heating the surface and applying various chemicals to cause a reaction to occur with the base metal of the casting—usually the copper, with some subtle influence from the other component elements that may be present. The goal is to create an attractive mineral-like coating on the surface. This coating is stable and affords some protection to the base metal, but generally speaking these surfaces are primarily esthetic and need further protection from the environment. Figure 3.39 shows a golden patina on the Henry Moore

FIGURE 3.39 *Two Piece Mirror Knife Edge* by Henry Moore, East Building, National Gallery of Art.

sculpture in front of the East Building of the National Gallery of Art in Washington, DC. The color and stippling on the surface of this beautiful patina is most likely developed from applying a ferric nitrate over a darkened base, then stippling and sanding with bismuth nitrate or possibly titanium oxide. It is the rich golden hue that has the indications of a ferric nitrate patina.

There are numerous patinas that can be applied to bronzes. They involve varying degrees of difficulty to apply and some are quite fragile. You can layer patinas to create enhanced effects to the surface. Mottling, highlighting, and polishing can bring out the luster of a semitransparent patina. Many foundries specialize in certain patinas and most maintain proprietary mixes and techniques that they use for the castings they produce. There are several very common patinas that have been successfully used for decades on bronze statuary. Figure 3.40 shows one of the more common patinas, French brown. This durable patina is well suited to the outdoors and can have subtle variations to give a unique attribute to the surface. This incredible casting is located in Miami.

There are numerous books dedicated to the patination of copper alloys for art. The various formulas and the characteristics of the patina are described. See the Further Reading section at the back of this book for several titles dedicated to patination presented for those interested in the art.

FIGURE 3.40 French brown patina on cast pillar with bas relief, by Manuel Carbonell.

TEXTURES

From earliest times after the discovery of copper, the ability to hammer the metal into shapes and forms was one of the most endearing properties of this special rock. The shaping of clay could be performed by hand then heated in fire to a hard, useful form. The shaping of copper, however, required not only heat but force. Early humans would have soon realized that unlike clay, copper would soften when heated, allowing further shaping. Shaping with force brought shape and stiffness. Hammering and heating repeated over and over allowed for the creation of forms no other material known at that time could match. If a copper part was damaged, or broken from use, it could be remelted and reformed. Copper had to be extremely valuable to early humans.

Plastic Deformation Surface Texturing

The ductility of copper alloys enables numerous techniques of surface shaping. These techniques have been used on copper and its alloys for centuries. Artistic surface techniques first used by early craftsmen have been transformed by modern methods of imparting designs into the malleable surface of copper alloys.

Repoussé and chasing are two techniques still in use today that have evolved over the decades to involve computer-aided design and manufacturing.

Repoussé and Chasing

Two early techniques that relied on the ductility of copper alloys were repoussé and chasing. These were performed on wrought material—usually sheet copper, or later, bronze or brass sheet. Early sheet forms were probably made by hammering the metal thin as it was heated, with care taken not to crack the metal as the work hardened.

Repoussé involves hammering copper from the reverse side to create a design in relief on the face side. Figure 3.41 shows several examples of this and the chasing metalworking technique. The French term "repoussé" means pushing up or out and refers to the technique of raising the metal by hammering on the reverse side. Chasing is hammering on the front surface and embossing a design into the copper. The term "chasing" is based on the French term *chasser*, which means to drive or push out. Chasing involves pushing the metal from the face side and moving the design onto the surface. Both are often performed on the same sheet of metal to create very intricate detail.

Special chasing hammers and tools are used to produce this delicate art form. In antiquity, wood and bone were the tools used. The art is usually drawn onto the copper surface and then outlined with scratching tools before the work is drawn by hammering.

The artist works the metal into a firm but soft and pliable supporting substance. Today carpet or various hard rubbers are sometimes used but originally the work was done using chaser's pitch. The pitch is warmed to make it flow into the metal as it is shaped. The pitch is gummy and adheres to the back of the metal and flows into the form. Pitch provides a firm backing that moves and flows as the metal is hammered.

Incredible detail can be achieved given the propensity of copper to shape and stretch and depending on the artist's skill. Repoussé is a learned skill and the process is slow and tedious. For architectural surfaces, modern shaping and tooling play a role in moving metal into established dies or hammering and stretching the metal.

Hammered Texture

Utilizing the softness of copper rather than the stiffer and stronger alloys, hammering textures is a step away from the artistic repoussé. The temper and work-hardening nature of copper needs to be well understood. The potential for inducing texture in copper alloys by means of pressure or hammering and adding color by means of patination or highlighting the surface is what affords

FIGURE 3.41 Repoussé and chasing work on ornamental panels.

copper alloys a further dimension in architectural surfacing. A simple texture can be created by hammering or rolling a sheet copper alloy material, and the sheet can then be darkened and used as surfacing material as shown in Figure 3.42. The metalwork hardens and becomes stiffer and harder as the cold working operation of the surface texturing creates the beautiful patterning.

Hammering is more of a surfacing technique rather than true repoussé work, but it lends itself to more elaborate design development for surfaces made of copper alloys.

The metal can be heated and hammered into forms or shapes without ripping or tearing along edges, then oxidized to develop beautiful tonal affects no other material can match. Figure 3.43 shows deep hammering on surfaces of copper with layering of one set of forming operations over another. This demonstrates the incredible ductility of the metal. Few materials offer this level of ductility coupled with beauty and contrast from the patina.

Embossing

Embossing is a process of running sheets or coils of thin copper alloy through engraved patterning rolls. The rolls apply pressure, which imparts the pattern into the copper alloy. Figure 3.44 shows

FIGURE 3.42 Hammered copper surface used on a kitchen hood in a residence.

an example of this. In this image the copper sheet has been embossed with a small hammered, bubble-like texture, then oxidized. The top and bottom images are of the surface with the texture coming out. The middle image shows the texture going in.

There are numerous textures that are rolled into copper alloys. Wood grain patterns, leather grain patterns, and stucco patterns are just a few. The softness of the metal leaves a texture with a defined edge. The texture stiffens the copper surface and conceals minor surface flaws.

There are other methods of embossing these soft copper alloys. One involves creating patterns in copper alloy by placing thin sheets under pressure over a rigid die. First the pattern is created in a rigid substrate, which acts as the die. The size and shape are limited by the press and the complexity of the form. Then pressure is applied on a rubber pad acting as a tool, which presses the copper alloy into the mold. The images shown in Figure 3.45 are examples of the patterns created by this method, has its roots in repoussé. The limitations of this technique are in the size of the die and design: the metal can rip if pressure is too great and if the metal is restricted from moving into the form.

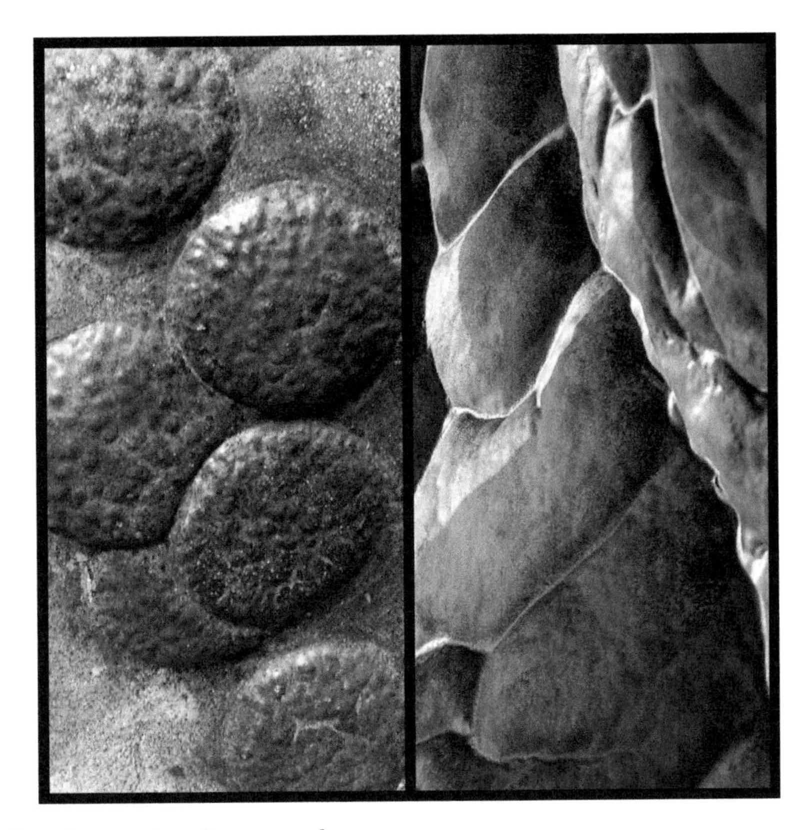

FIGURE 3.43 Deep hammering of copper surfaces.

Computer Numerical Control Texturing

With the advent of modern CNC equipment, the ability to shape and form copper alloys has advanced in unique and intriguing ways. This is allowing one-of-a-kind textural surfaces to be introduced into copper and copper alloy surfaces. Figure 3.46 shows examples of textures induced into copper to produce tactile surfaces that would be used either in an interior or exterior application.

Larger textures are also possible using CNC equipment. These macro forms in the metal produce shadowing effects that will oxidize at different rates depending on how the condensation remains on the surface. It may be a stretch to relate the action of a CNC press to the ancient practice of repoussé, but the way the metal is locally stretched to develop artistic expression is similar. It is the ductility of the copper alloys that affords this behavior. The metal is held and plastically deformed only in a select area. Other metals can also be shaped in similar fashion, but copper and many of its alloys do not develop excessive local stresses that can warp the surface. The metal does work harden and there are limits to the elongation of the metal; still, remarkable surface shapes can be achieved. Figure 3.47 shows the surface of the de Young museum's copper walls reflecting off the glass in the

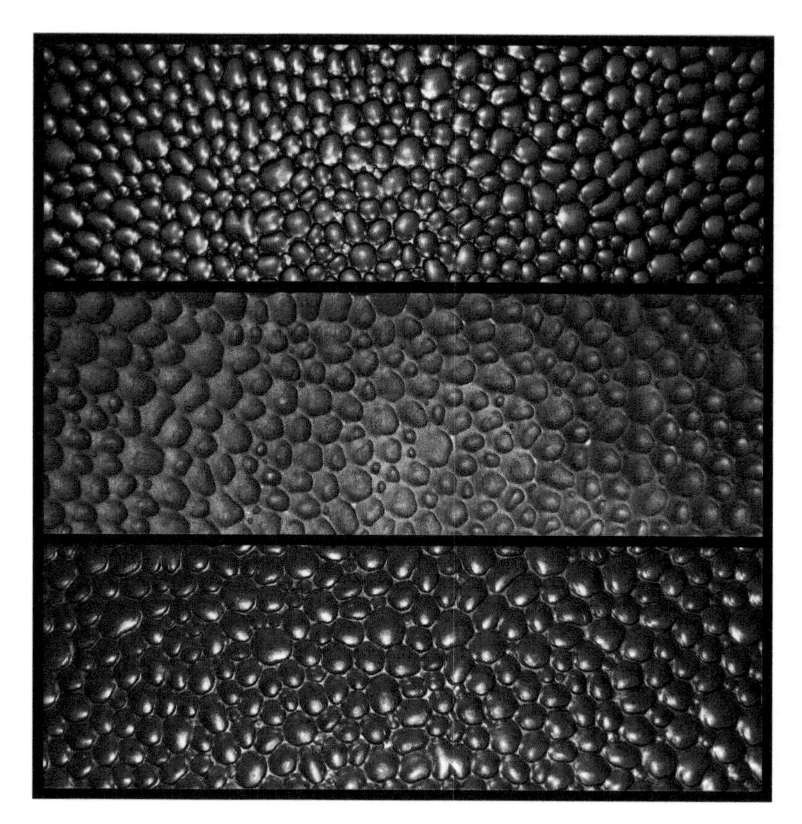

FIGURE 3.44 Embossed copper.

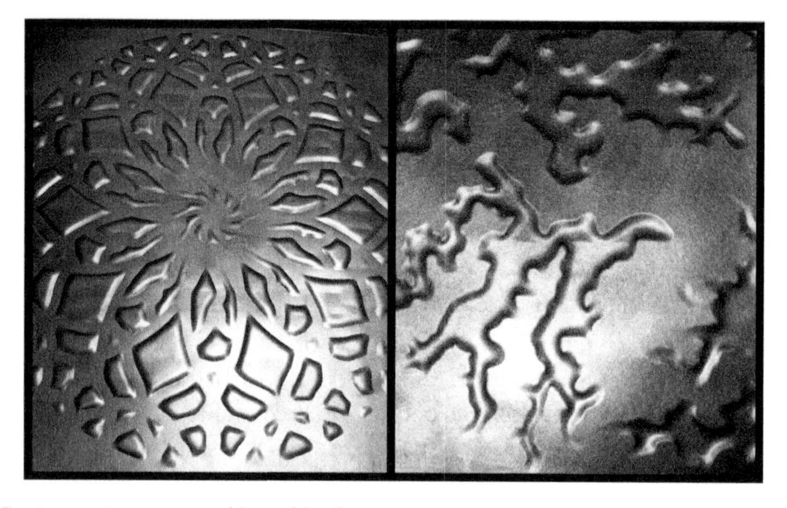

FIGURE 3.45 Custom patterns pressed into thin sheet copper.

FIGURE 3.46 Tactile surface textures produced using CNC.

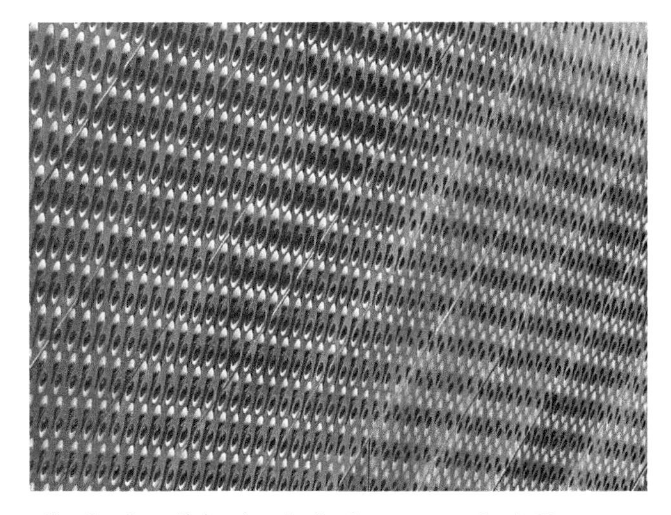

FIGURE 3.47 Copper wall reflecting off the glass in the fern court at the de Young museum in San Francisco, Source: by Herzog & de Meuron.

fern court. Here a series of nine different cone-shaped indentations were pushed inward or outward from the plane of a sheet to develop subtle contrast on the museum's exterior walls.

Etching and Engraving

Copper and copper alloys can be etched or engraved to produce extensive detail in the surface. In both processes material is selectively removed. Etching involves localized chemical dissolution of the metal and usually necessitates application of a resist to the surface of the metal not to be dissolved.

The etched region can also be treated to produce a contrasting design. The chemical etched surface is very receptive to reacting with other chemistry to form a patina. One benefit to this is that the masking where the metal was dissolved is already in place. Once the etching has occurred, the plate can be rinsed then immersed in a darkening solution to create highlights or, in the case of the work shown in Figure 3.48, the surface can be treated to create color tones in the copper alloy. This artwork was created using the C26000 alloy, which then had a statuary finish applied to darken it. The orange and red colors are produced by dyeing the copper alloy surface.

Engraving removes metal by mechanical means. Selective abrading, machining, and waterjet and laser cutting can engrave the metal's surface. A resist is not necessary, but fine machining and control of the tool is critical.

Figure 3.49 shows examples of several methods of etching and engraving. The image at the top left is a close-up view of a C22000 alloy plate that has been chemically etched. The low parts of the

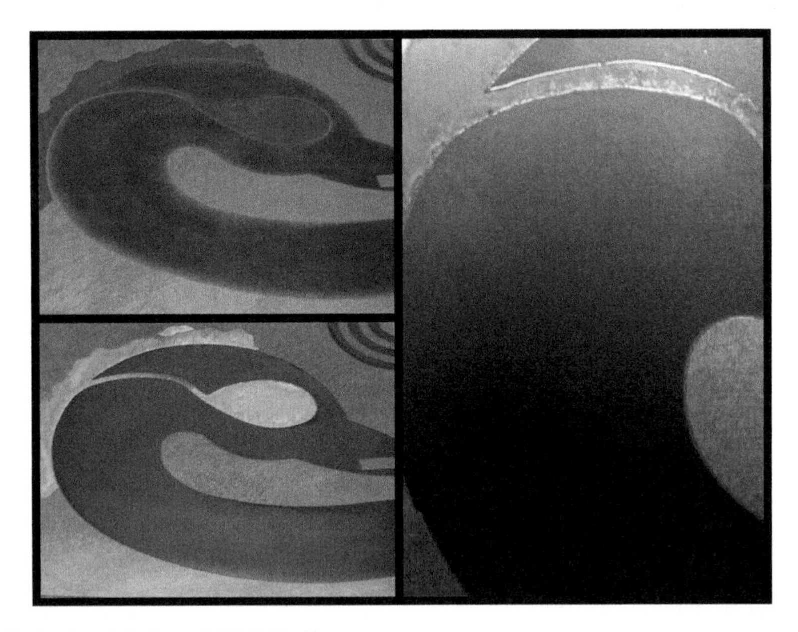

FIGURE 3.48 Etched and darkened C26000 alloy.

FIGURE 3.49 Various examples of etched and engraved copper alloys.

etched surface were darkened and the top sanded and hammered to develop a contrasting relief. The image at the top right shows a C26000 alloy plate that was machined to produce an image of a leaf then leafed with copper and silver leaf. The image at the bottom shows textures that were engraved using waterjet cutting (left) and laser cutting (right).

Chemical etching of copper has been around for centuries. Also referred to as "chemical milling," bronze work, helmets, shields, and swords were etched by applying pitch or tar to the surface, scratching the artistic design into it, then exposing the piece to various mineral acids. Where the scratches exposed the metal, the acid would dissolve small amounts. The pitch was removed, and the design was revealed in the metal surface.

Today, various etching solutions are used on copper alloys. Most common is ferric chloride, but cupric chloride is also used. The resists are created by applying a light-sensitive film and exposing the film to intensive light while the image to be etched is overlaid. Once exposed, the light-sensitive film hardens and the balance is rinsed from the surface, leaving the metal exposed. The exposed metal has the etchant applied and this dissolves some of the metal, creating an image. Once the image is etched into the surface, the resist is removed (Figure 3.50).

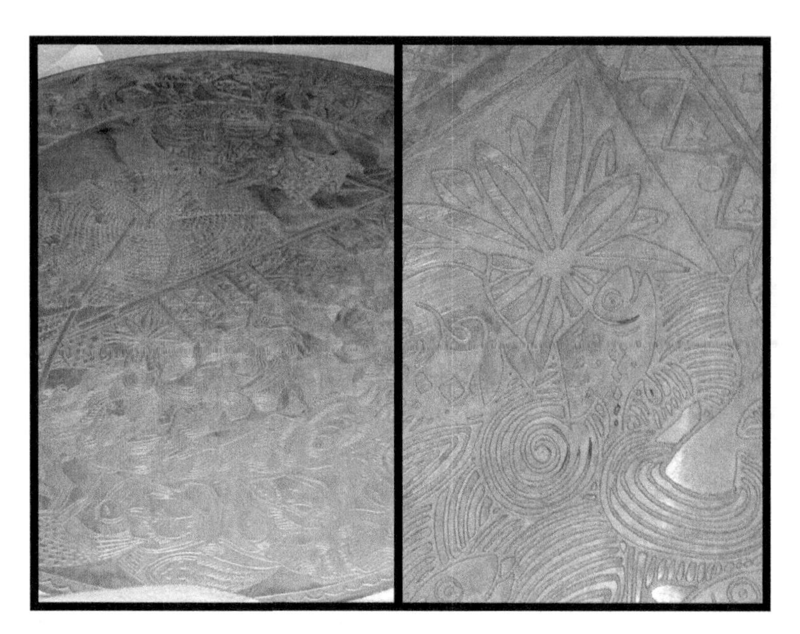

FIGURE 3.50 Chemical milled C22000 plate below.

Engraving removes metal by mechanical means in the sense that waterjet systems and laser systems can be "tuned" to only affect the metal up to a given distance into the surface. This approach can take a drawing and digitize it for the CNC operation of the two axis machines. The system uses metal ablation to remove material from the surface to a depth of around 1/10 of the thickness of the material. The groove created can then be filled with enamel or darkened chemically while the surface is finished to create a contrasting effect. Lists of names, designs, and other artistic expressions can be introduced into the surface.

Engraving can also involve controlled scratching of the surface. This is the way the work was created during the Middle Ages and in antiquity, but modern CNC systems can speed this process up and produce designs directly from a digital drawing. This is also a method of cutting the resist to enable chemical etching or milling to occur.

The top image in Figure 3.51 shows the C22000 alloy with the names of the victims of the 9/11 attack at the World Trade Center. These were cut into the copper alloy using laser etching. The lower left image shows a design produced by waterjet cutting of C75200 nickel silver. The image at the lower right is an engraved design created by scratching the aluminum bronze (C61500) surface with a special bit controlled by CNC on a machining center.

Plating

Copper alloys are particularly adapted to plating. Plating consists of the deposition of a thin layer of metal onto the surface of copper alloys. There are two methods in common use: electroplating and electroless plating.

FIGURE 3.51 *Top*: Laser-etched names in Commercial Bronze (alloy C22000). *Lower left*: Waterjet-engraved design in the nickel–silver alloy C75200. *Lower right*: Engraved design in the aluminum–bronze alloy C61500.

Electroplating requires that an electrical current be established, as the copper alloy is submerged in an electrolyte containing ions of another metal. The ions deposit evenly across the surface as the current is applied. You can selectively electroplate copper, also referred to as brush plating, using resists and a special transfer wand or brush to apply the electrolyte and establish a localized current.

Electroless plating involves immersion of the copper alloy into a solution containing several chemicals that create a reaction, after which an autocatalytic process occurs. When this occurs, hydrogen is released from the surface of the copper alloy, causing a negative charge to develop. The metal ions in the solution are positively charged and they deposit on the copper surface.

The left-hand image in Figure 3.52 shows silver selectively plated on a copper surface. The right-hand image shows a large casting with a chrome-plated "soccer ball." The ball was chromed separately and then attached to the casting.

For either of these processes, the copper alloy surface must be very clean and free of oxides, oils, and fingerprints. These will inhibit the deposition process by interfering with the electrical charge and the current flow. Metals often used to deposit on copper alloys are nickel, tin, silver, chromium, and even gold.

The conductivity of the copper alloy makes them very effective for plating, and intricate designs can be created using resists and selective plating methods. The top image in Figure 3.53 shows an art piece designed by Jan Hendrix made from C22000, which was chrome plated to produce a mirror surface on the copper alloy. All edges and surfaces received the mirror-like chrome. The image at the lower left is a nickel-plated C26000 floor plate that has been etched and filled. Here, only

FIGURE 3.52 *Left*: Silver plate on copper demonstration sample. *Right*: Chrome plate on silicon bronze.

FIGURE 3.53 *Top*: Chrome-plated *Fuga*, by Jan Hendrix. *Lower left*: Nickel plate edge on a plaque. *Lower right*: Silver plate on copper.

the outer perimeter was plated in nickel. The image at the lower right is a silver-plated design on the C11000 alloy.

When correctly performed, plating is very durable method of creating color tones and inlay effects on copper alloys. The deposition is relatively fast due to the conductivity of copper and the result is a very stable metal-on-metal coating.

TIN-COATED COPPER

Tin coatings are deposited onto clean copper sheet for producing a gray, thin roofing and siding material. There are companies that regularly produce tin coatings on copper for industrial use and for architectural use. In the image on the left in Figure 3.54, a thin, tin-coated copper sheet is used as roofing; the right-hand image shows the same metal after a few years of exposure.

Tin-coated copper was introduced as a replacement for lead-coated copper roofing and sheathing. The tin is nontoxic and lacks the stigma of lead surfaces. Additionally, many lead-coated copper surfaces age in peculiar ways in today's sulfur-free environments. In the past, lead-coated copper would develop a dark lead sulfide on the surface, but today the lead develops a streaky surface of lead oxide that can be red or white, leaving an unpleasing appearance.

FIGURE 3.54 Tin-coated copper.
Source: Photo courtesy of KME.

MELTED COPPER ALLOY SURFACING

Copper alloys have different melting temperatures as the constituents that make up the alloys melt at different rates. As the surface is heated to the point where the metal reaches its liquidus temperature (the point at which it begins to melt), some of the components vaporize while others move about in the molten metal on the surface. Copper alloys move the heat away quickly. As you blast the surface to bring the temperature up, the copper and other components flow. The effect is like the activity on the surface of Venus: molten metal flows, then solidifies. Each alloy heats and cools differently and this creates some remarkable surface affects. Figure 3.55 shows several examples of this phenomenon. The top-left image shows the alloy C75200 (nickel silver). The middle image in the top row shows C22000, while the top-right image shows the alloy C28000, or Muntz Metal. The image at the bottom shows the large plate forms of the C46400 alloy that comprise an art piece produced for the 9/11 Memorial in Overland Park, Kansas.

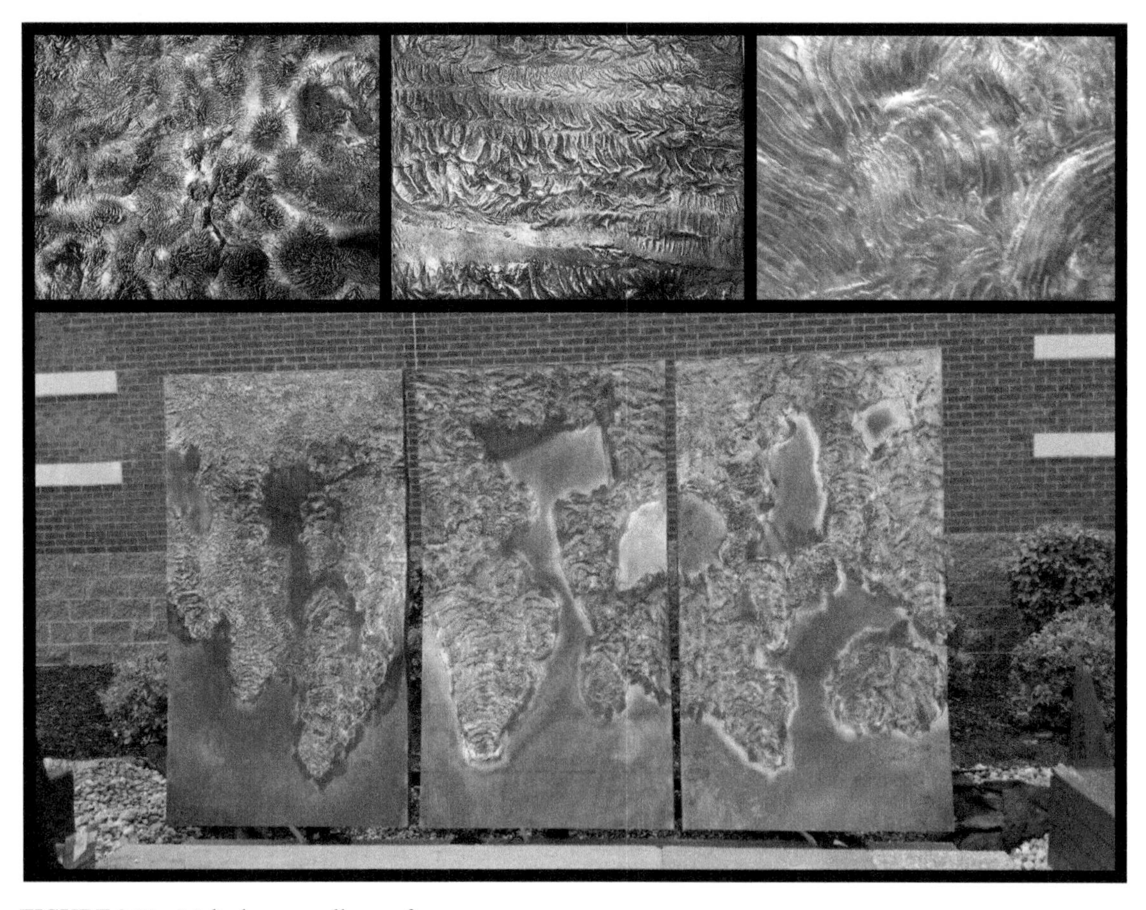

FIGURE 3.55 Melted copper alloy surfaces.

It goes without saying that working with this technique requires respiratory equipment as well as heat protection. Some shaping occurs as the metal heats and cools differentially.

COPPER AND GLASS

Another interesting phenomenon occurs when combining copper and glass. Copper and glass have different thermal coefficients of expansion. Therefore, when joined together you would expect the glass to crack as the copper moves. Experiments in which the glass is joined with the copper by heating both to a temperature at or near the melting point results in the copper annealing, making it very weak and soft. Figure 3.56 shows works of copper and glass created by two different methods. The image at the top shows a work created by layering several plates of glass with several copper

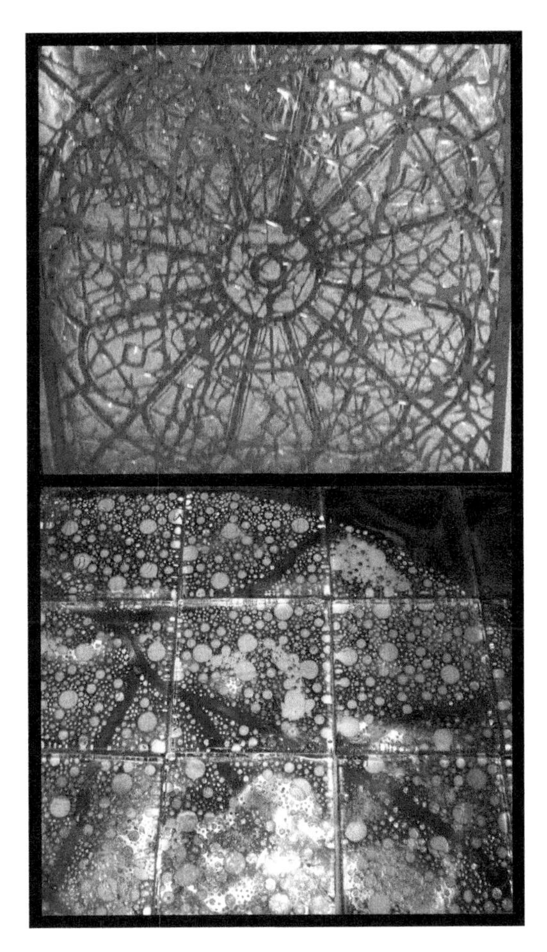

FIGURE 3.56 Copper and glass.

sheets, then using waterjet cutting to create a design. The work in the image at the bottom was made by taking molten glass and pouring it over copper sheet that has been cut out in a leaf design.

The intense red color in both works was created by first treating the copper to bring out the cuprite. As the glass melts during processing, it locks the cuprite into the surface in a deep, blood-red color. Both of these works have been exposed to mild changes in temperatures for several years.

PROTECTING THE SURFACE

Protective coatings are closely associated with copper alloys used in art and architecture. The protective coatings are clear and contribute a barrier of separation from moisture, pollutants, and other potentially damaging substances. There are tinted waxes and lacquers that sometimes are used as coatings for copper alloys. These are used to cover or emulate a statuary finish. In this case they are not much different from a paint, and you will lose the real metallic nature of the metal.

If the intent is for a copper alloy surface to remain unchanged, it should be protected from the ambient environment, whether interior or exterior. Protective coatings include the addition of natural color tones with a finish or polish, statuary oxide finishes, and even patinas. One of the most common protective coatings used to temporarily protect the surface is wax. But even the best hard waxes will not stand the test of time; they dry out and degrade over time. Waxes must be removed and reapplied periodically, depending on the exposure. For exterior statuary new wax may be needed yearly, depending on the climate and exposure. Lacquers are another common coating used to protect the surface of copper alloys. Lacquers can be coated with wax to add further protection to the surface.

Chapters 4 and 8 go into more depth on protective measures, the characteristics of various coatings applied to copper alloys, and decorative patinas.

CHAPTER 4

Expectations of the Visual Surface

To practice the art of patination is to collaborate with nature and time.

Ron Young, *Contemporary Patination*

INTRODUCTION

Copper alloys are used in art and architecture because of their beauty and color. They have an elegance and panache that reflect a yearning as old as when mankind first gathered up the beautiful rich green malachite or crystalline deep blue azurite minerals.

Copper has the weight of steel and feel of silver. In its unalloyed form it is more ductile than aluminum or zinc, yet when alloyed with other elements it can possess good strength and hardness. The resistance to destructive corrosion in our environment, whether rural, urban, or seaside exposure, is a major strong point of the metal. In reality, much of what we appreciate about the metal, at least for natural copper surfaces, is due to its interaction with the elements in our immediate environment. The pleasing green patinas that adorn the buildings built more than a century ago give copper a distinctive prominence in our impression of architecture.

Surface appearance and durability are important factors today in the decision to use copper in art and architecture. The color and patinas that form play a significant role in this decision. The exposure of the copper and copper alloy surface involves interaction with the selected environment. This is the case whether the surface is prepatinated, coated with an organic film, or left to change naturally as the surface ages. The images in Figure 4.1 show the intense, but short-lived, color of newly installed copper.

When exposed to the atmosphere, copper readily combines with oxygen to form one of its oxides. As shown previously in Figure 1.2, a copper atom has one valence electron in its outer shell. This allows electron sharing with other atoms, such as oxygen.

FIGURE 4.1 Newly installed copper on a roof in Minnesota and on walls in Texas.

When moisture is present, the free electron in the outer shell bonds with oxygen. If only one copper atom bonds, the compound cupric oxide (CuO) forms. This compound is characterized by a dark, sometimes black, color. If, however, two atoms of copper are involved in bonding, the compound cuprous oxide (Cu_2O) forms. This compound is a reddish color and is typical of what forms initially on copper. Figure 4.2 shows the development of cuprous oxide on the surface of the copper panels used on the de Young museum in Golden Gate Park, San Francisco.

Over the decades, many tests have been performed to evaluate how copper and copper alloys perform in atmospheric exposures when left unprotected by films or predeveloped oxides. In Sweden, corrosion tests were performed on 36 different copper alloys in sheet and in rod forms.[1] The tests were performed in rural, marine, and industrial locations. Another series of tests was performed

[1]R. Holm and E. Mattson, "Atmospheric Corrosion Tests of Copper and Copper Alloys in Sweden—16 Year Results," in *Atmospheric Corrosion of Metals*, eds. S. Dean and E. Rhea (West Conshohocken, PA: ASTM International, 1982), 85–104.

FIGURE 4.2 Cuprous oxide forming on the C11000 panels used at the de Young museum after 10 years of exposure near the ocean.
Source: Architects: Herzog & de Meuron.

by a separate study in the United States on several copper alloys in common use.[2] This study was performed in rural, marine, and industrial locations as well. In both studies, corrosion of copper stabilized after a period of a few years, with the copper alloys forming the initial deep oxide on the surface in the first two years.

Both studies found similar corrosion rates, with the lowest corrosion rates in rural exposures as expected and the highest rates found in industrial exposures. The industrial sites also developed the richest patina of green on high-copper alloys.

Location	Rate of corrosion ($\mu m\,yr^{-1}$)
Rural	0.3–0.7
Marine	0.5–0.9
Industrial	0.9–1.4

[2] L. P. Costas, "Atmospheric Corrosion of Copper Alloys Exposed for 15 to 20 years," in *Atmospheric Corrosion of Metals*, eds. S. Dean and E. Rhea (West Conshohocken, PA: ASTM International, 1982), 106–115.

These studies indicate that the rate of corrosion of copper and copper alloys exposed in various atmospheres is extremely low. Little to no pitting was observed in general, and only in the high-zinc alloys; alloys with the alpha-beta phase, such as C26000, showed extensive signs of dezincification corrosion.

Patinas formed on all the alloys, but the form and color of the patina varied with the exposure and with the copper alloy makeup. In each of the studies the surface showed a marked darkening in the first two years. In the commercially pure copper alloy C11000, the surface turned dark brown

FIGURE 4.3 Interference oxides developing on copper surfaces exposed to the atmosphere after a few weeks.

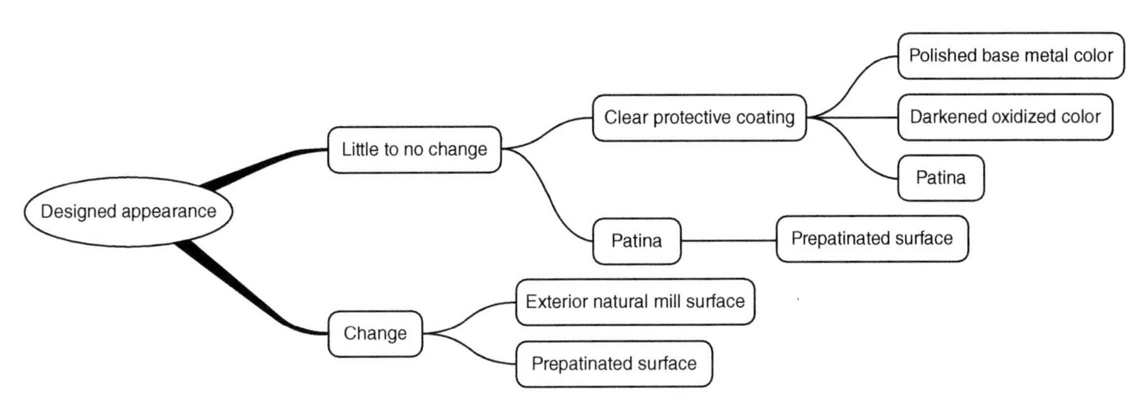

FIGURE 4.4 Distinct design paths for the use of copper alloys in art and architecture.

to black in rural and industrial locations and brown in marine locations. This would correspond to the development of cupric oxide, the transition film on copper alloys.

After 16-plus years of exposure, the copper developed a uniform brown-green to green color in the industrial and marine locations and a blackish brown shade in the rural exposures. Brass alloys in rural exposures turned black, while surfaces in marine exposures turned a dark grayish green tone. Brass alloys turned a brownish green tone in urban environments.

As the copper surface is exposed to humid atmospheres, the thin oxide begins to form. The copper still retains some of its initial reflectivity, and this coupled with the oxide develops thin-film interference as shown in Figure 4.3. This effect is short lived, maybe enduring for a few months, because as the oxide grows the interference effect diminishes.

There are two distinctively different paths a design will take when copper and its alloys are used: one path envisions a metal surface that will remain unchanged, while the other anticipates a metal surface that will be allowed to change and weather naturally. Figure 4.4 shows these two paths and their potential branching.

INTENT: AN UNCHANGED SURFACE APPEARANCE

For some designers, one of the biggest drawbacks of using copper and copper alloys is the entropic drive the surface undergoes. Copper surfaces seek to combine with oxygen and other substances, which alters their reflectivity and color. This creates contrasting light and dark regions, spots, streaks, and color changes to an otherwise jewel-like surface.

Producing patinas and statuary finishes on copper alloy surfaces slows the changes in appearance and can sometimes mask these changes, at least for a period of time. The chemical reactions these processes involve slow any additional oxidizing tendency; however, eventually they too will be affected by the environmental exposure. Designs in which the natural or refined surface color is intended to remain untouched by further oxidation pose the greatest challenge.

This path can involve a polished, natural-finish metal surface or an oxidized and chemically enhanced surface. Many light and medium statuary finishes obtain their beauty by creating an enhanced oxidized surface with a level of transparency, allowing the underlying metal to show through. These semitransparent finishes act similarly to natural finishes; in a way they have been pre-aged and this adds a small amount of corrosion protection to the metal surface.

The need to limit the oxidation behavior of copper and copper alloys is found more often when they are used in decorative interior surfaces or elegant exterior entryways. These areas are subjected to different humidity levels and to more frequent contact with hands and the oils of perspiration. Copper and its main alloys combine readily with oxygen and oxygen compounds, such as carbonates and sulfates. These will darken the surface if not sealed. Fingerprinting also presents this issue, as copper alloy surfaces will quickly react with the oils and other substances in human perspiration and form tenacious oxides, almost as if they have been etched into the surface.

Surface Preparation

To thwart the entropic drive of the surface oxidation process, various protective measures have been utilized. These measures involve a protective coating. Results from these protective measures depend a great deal on the surface preparation, application of the coating, type of coating, and surface maintenance after exposure.

Keys to protective coating results:

- Surface preparation
- Coating application
- Coating type
- Maintenance program

There are two parts to surface preparation. Surface preparation always begins with a clean copper alloy surface. Whether the surface is a mill finish, mechanically polished, or chemically oxidized, the surface must be clean and free of all oils, polishing residue, fingerprints, and other foreign substances that may interfere with the soundness of the coating or cause the metal to react over time.

In preparing the copper alloy surface prior to the application of a protective coating, it is necessary to arrive at a very clean oil- and grease-free surface. There should be no fingerprint residue whatsoever and no moisture. The removal of moisture, oils, and polishing residue is paramount. Wiping the surface with an alkaline degreaser and 99% isopropyl alcohol is advised. As you work with the metal, wear clean cotton gloves to avoid fingerprint oils. Keep the surface dry and free of foreign substances.

In the second part of the surface preparation, an oxide inhibitor is applied. This involves wiping or dipping the metal fabrication, sheet, plate, or extrusion with a solution of benzotriazole. The solution can be a 2% solution of benzotriazole and distilled water or benzotriazole with alcohol,

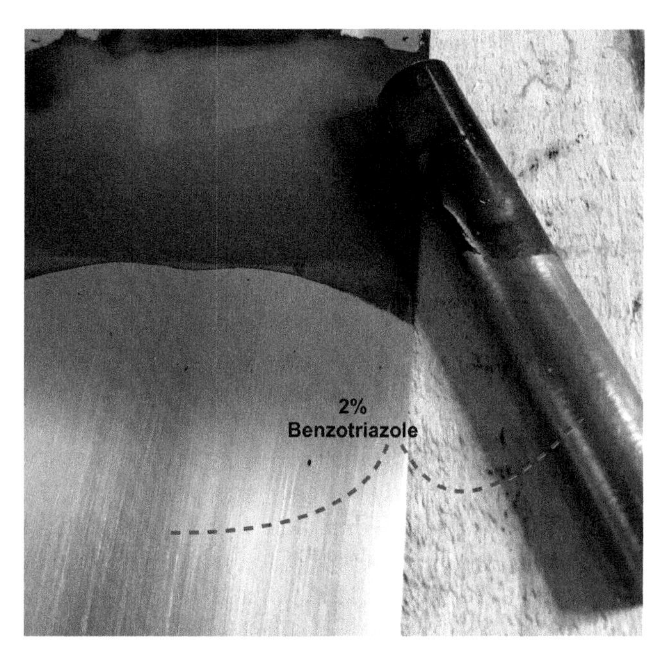

FIGURE 4.5 Benzotriazole test.

such as 99% isopropyl. This inhibitor will prevent the surface from oxidizing as further work is performed. Another method that is found to be successful is to dip the copper alloy into a solution of potassium ferrocyanide and 2% benzotriazole for 1–10 minutes. This treatment will keep the surface from oxidizing.

Figure 4.5 shows a copper sheet and copper tube that have been partially submerged in a 2% benzotriazole and alcohol solution and allowed to dry. Then fully submerged in a strong blackening solution of selenious acid and hydrochloric acid. The portion of the tube coated with the benzotriazole remained untouched by the darkening solution.

Coating and Application

Once the surface is cleaned and coated with an inhibitor and all forming and shaping has occurred, it is ready for the application of a clear protective coating. Note that you may need to wipe the surface again with the oxide inhibitor to be sure subsequent activity did not alter the protective treatment.

Specially formulated clear acrylic lacquers have proven over time to be the best protection for the surface of copper alloys. In the art restoration world, many conservators shy away from these coatings and prefer high-grade waxes when treating sculpture. For architectural surfaces, clear acrylic coatings are often used. There are nano-coatings in development as well.

The purpose of these coatings is to seal the copper alloy surface from moisture and oxygen. One of the biggest drivers of corrosive action on metals is relative humidity. As the ambient temperature changes, metals lag behind and remain cooler longer, which causes condensation to develop on the surface even in the driest of regions. This condensation often collects substances that possess an ionic charge when in solution. These can seek out the smallest breach through the coating and work to combine with the metal. Eliminating or reducing this porosity is the goal of a successful application.

Incralac, an acrylic lacquer developed from research sponsored by the International Copper Research Association (INCRA), is an excellent coating for copper alloy fabrications. The coating has been around since the late 1960s. This coating has benzotriazole along with ultraviolet light absorbers and chelating agents to improve coating consistency and inhibit under-film corrosion. The acrylic resin is the PARALOID B-44™, manufactured by Dow. It possesses good hardness, flexibility, and adhesion. It is a softer acrylic than most other acrylics and urethanes. Incralac is formulated with 17% solids and a blend of solvents, mostly toluene and xylene.

For cast bronze and brass statuaries, many conservators prefer the application of a hard wax over the lacquer. They feel wax is easier to remove and replace, while lacquer requires more hazardous solvents, such as toluene or xylene, to remove it. Additionally, lacquers can leave the sculpture with a glossy appearance.

Cast sculptures placed outdoors are exposed to the elements, including rain and humidity, as well as bird waste and, at times, human contact. Cast sculptures placed in fountain settings have the added detriment of the minerals in the water collecting on their surfaces. Some exposures of outdoor sculpture, as well as building entryways, involve occasional contact with deicing salts. If the coatings are not maintained, exposure to these conditions and substances will cause oxides to develop. The oxides can show up dark mottled tones on the more reflective backgrounds, or they can develop copper salts with contrasting steaks of color. Figure 4.6 shows a sculpture exposed for years in a fountain. Each winter the fountain was turned off, but the sculpture was never maintained. The original patina has been attacked and any wax protection has long since disappeared.

These oxide formations initially only harm the patinas that gave the sculpture color. They can, however, eventually etch the surface of copper alloys if they react with the patina and begin to join with the base metal. They will alter the esthetics of the surface by changing the original gloss and color. The length of time of exposure and type of exposure will determine the difficulty of restoring the surface.

If the patina has reacted and compounds have developed, removal of the contrasting oxide will remove the patina as well. Patinas are themselves oxides and mineral substances. The new oxide growth is an extension of the original patina and cannot be removed without also removing the patina.

Lacquers offer greater durability and have longer working life spans than waxes. Waxes, even the most durable waxes, require more frequent maintenance and renewal. They must be removed and replaced on a regular basis. Combining lacquers and waxes is a way to achieve the benefits of both. Applying a sound lacquer, such as Incralac, and then applying a good, hard wax over the

FIGURE 4.6 Fireman's Memorial fountain sculpture. Designed by artist Tom Corbin.

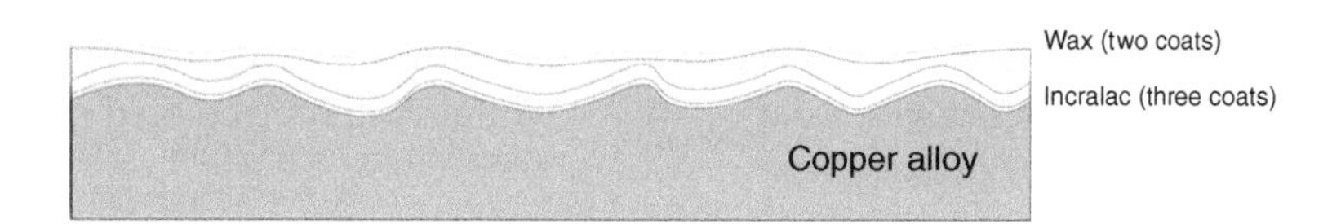

FIGURE 4.7 Sketch of Incralac lacquer and a wax coating on a copper alloy.

surface can extend the life of the lacquer and thus protect the sculpture's surface more effectively (Figure 4.7).

The effectiveness of a lacquer-wax coating is achieved by reducing the potential for porosity through the coating, by the layering of the benzotriazole, and by the ultraviolent absorbing agents. The wax gives additional ultraviolet protection, takes down the glossy sheen slightly, and adds weathering protection to the lacquer. The wax provides hydrophobicity to the outer surface. The wax will adhere to the lacquer. If using heat to assist in applying the wax, the heat should be kept low to avoid damaging the lacquer.

TABLE 4.1 Coating comparisons.

	Wax	Incralac	Incralac and wax
Life span	3 years	10–12 years	15–20 years on lacquer
Gloss	Low	High	Medium
Application	Brush	Spray	Brush
Ease of removal	Simple	Difficult	Simple
Solvent	Mineral spirits	Xylene	Mineral spirits

As the wax degrades, it can easily be removed using mild solvents such as mineral spirits, which will not harm or remove the Incralac. New wax can be reapplied over the lacquer, further extending the protection and life span of the lacquer. When removing the old wax and before applying the new wax, any damage to the lacquer can be repaired. Incralac allows for touching up and blending older lacquer with new (Table 4.1).

For any copper alloy sculpture protected by any coating, the level of maintenance is key to sculpture's appearance. Newly waxed sculptures are stunningly beautiful. If the sculpture is not cleaned and the wax is not replaced, the surface of the sculpture will become drab as the degrading wax yellows. Waxes are made from hydrocarbons, which are plant, animal, or synthetic long chains of aliphatic hydrocarbons. Those used on copper alloy sculpture provide a hard, protective layer. Usually either carnauba, microcrystalline, or a combination of both is used to seal the surface of a sculpture and provide a hydrophobic layer. As the sculpture undergoes temperature changes and absorbs ultraviolet radiation, these long-chain molecules begin to break down and the protective surface layer develops cracks. Eventually the wax will degrade to the point where it disintegrates and sloughs from the surface, leaving the patina exposed to the weather. Figure 4.8 shows various copper alloy surfaces with wax coatings that have aged over time and exposure. The old wax can be seen as a white smear on dark patinas, while on lighter-colored patinas the color is variable.

Before this occurs, the old wax should be removed and new wax applied. You do not want to apply new wax over old wax. This leads to build up, and the underlying old wax lacks the needed flexibility to expand and contract with temperature changes.

The decision to use wax coatings or lacquer coatings boils down to whether or not the surface will be maintained on a regular basis. It is also important to decide whether the patina will be protected or allowed to undergo changes induced by weathering. If the sculpture is not going to have a regular maintenance schedule, then waxing will not suffice. Wax left to decay will not protect the patina long term, and the patina surface will undergo weathering.

A good lacquer coating, one with benzotriazole or tolyltriazole, will afford longer periods of time between maintenance intervals. When properly applied, these coatings will afford protection of the patina and the base copper alloy sculpture.

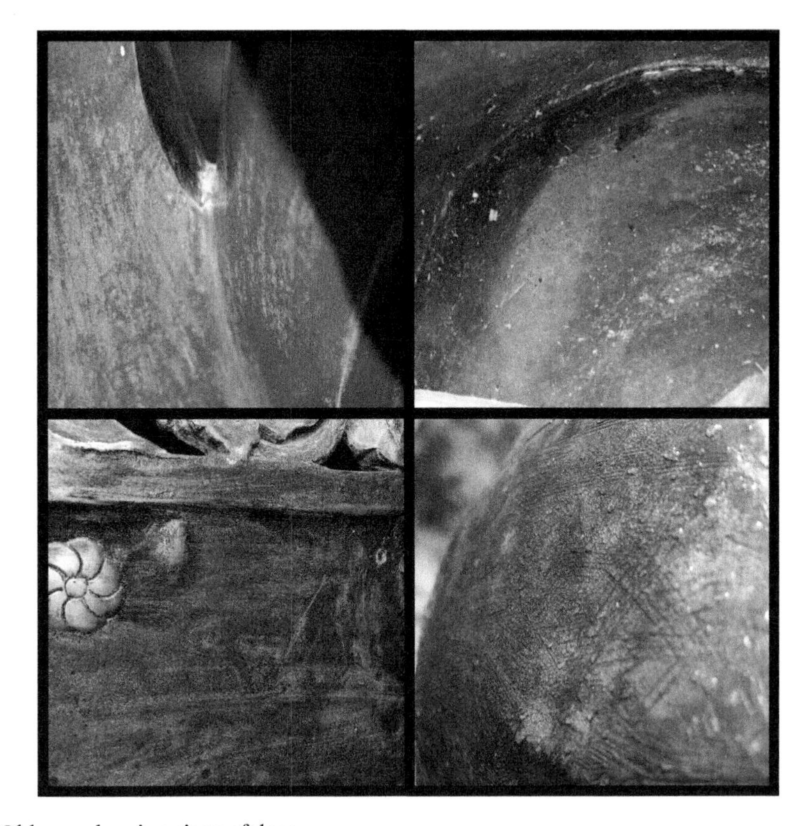

FIGURE 4.8 Old wax showing signs of decay.

For bronze sculptures, the patinas used vary in durability. They are chemical oxides formed by the reaction on the copper alloy surface, but they will react with the environment and will develop other mineral forms that will affect the base patina. These art forms should be coated and maintained to provide a surface close to the original richness provided by the artist.

Figure 4.9 shows a beautiful sculpture called the *Muse of Missouri* designed by the artist Wheeler Williams and erected in 1963. The wax was mostly nonexistent after a decade or so of exposure, and the environment was beginning to affect the patina, which was developing streaks of corrosion products on the surface.

For most brass surfaces, particularly those that have a bright surface appearance, coatings with clear lacquers or regular waxing is necessary if the surface is to maintain its luster. The finish on oxidized brasses, referred to as statuary bronze or oil-rubbed bronze, holds up better but still will succumb to humidity or the chemistry in human perspiration. This can be difficult with such things as handrails. Constant human touching will wear away any protective coatings and more often than not, leave the surface bright. Figure 4.10 shows images of copper alloy handrails in which the coating and the patina have been rubbed down to the bare metal by the constant handling.

FIGURE 4.9 The *Muse of Missouri* and stains developing on its patina.

FIGURE 4.10 Copper alloy handrails showing signs of patina wear from human interaction.

The Oxide Inhibitor Benzotriazole

There are several clear coatings in use as protective barriers for copper alloys. The coatings that incorporate benzotriazole will perform the best. Benzotriazole, sometimes referred by the letters BTA, provides a level of active tarnish inhibition to the copper surface. Benzotriazole can be incorporated into a lacquer, as in the case of Incralac, a coating developed specifically for copper alloys, or it can be incorporated as a light oil. The Mill producers of copper alloys use a light spray of benzotriazole oil to protect the surface of copper alloys during processing and storage. When correctly incorporated into the coatings and when layered, it has proven to show very good results. The chemical formula of benzotriazole is $C_6H_5N_3$. The unique attribute of this very thin molecule is that it can act as both a mild acid and a mild base. It is a polar molecule and can bind with copper on the surface, effectively inhibiting oxidation (Figure 4.11). It will partially dissolve in water, but when bound to copper the resulting molecule is insoluble.

Benzotriazole attaches to the copper and spreads out over the surface. When benzotriazole is incorporated into a lacquer coating it disperses over the surface of the copper alloy and aids in keeping oxygen from reaching the copper. Incralac is a lacquer based on PARALOID B-44 thermoplastic resin. This acrylic resin is durable and was developed specifically for copper and bronze. When used, Incralac is dissolved in a matrix of solvents, toluene, and xylene, typically at a level of 17% solids to 83% solvent blend. An ultraviolet light inhibiter is incorporated along with the benzotriazole.

For protecting outdoor surfaces, applying the coating in layers and allowing it to dry between each coating is recommended. As an added protective step, add a hard wax such as carnauba over the top of the Incralac layers to further protect the surface. All coatings have some measure of porosity.

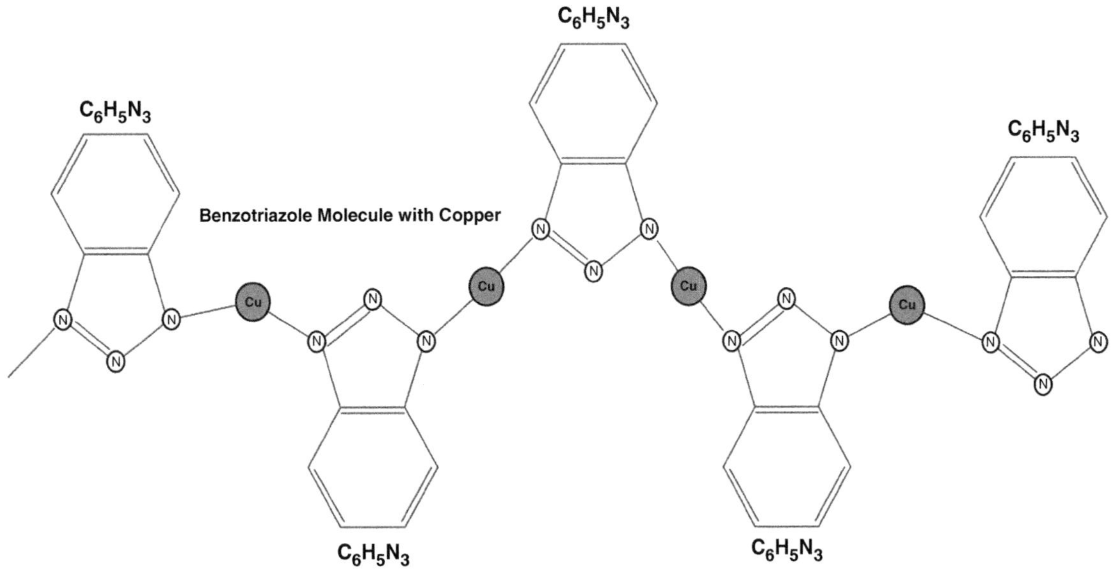

FIGURE 4.11 Benzotriazole bond with surface copper.

FIGURE 4.12 *Left*: *Heartland Harvest*, an artwork using the C11000 alloy at the Kansas City Board of Trade, after 24 years exposure. *Right*: C22000 statuary bronze panels at the Walker Art Building at Bowdoin College after 10 years exposure.

By layering the coatings, the porosity potential is reduced. Additionally, the wax protects the Incralac coating and can be removed and replaced as needed. To remove it, use mineral spirits, which will dissolve wax without dissolving the Incralac. Adding a wax layer over the lacquer coating is effective for sculptures and other small objects but not for large surfaces because the waxing and rewaxing can be costly. In these instances, where the surface area is large and a wax overlay is impractical, applying the Incralac or similar protective layers is critically important. Figure 4.12 shows two projects where Incralac was applied. The left-hand image shows a C11000 clear and patinated copper art wall after 24 years of exposure. You can see oxidation beginning to appear on the polished copper surfaces. The right-hand image shows an entryway of C22000 with a statuary bronze finish after 10 years exposure.

As mentioned previously, the durability and life span of these coatings is highly dependent on the initial preparation of the surface. The copper surface needs to be thoroughly cleaned of all oils and grease. Fingerprints will show up even after coating the surface if the metal is handled with bare hands or gloves contaminated with perspiration. Perspiration is composed of organic oils and fats called lipids, which are amino acids and water produced by the body, and there are often salts intermixed in the oils. The amount of chloride in a fingerprint is insignificant for the most part, but when trapped under the lacquer coatings it will continue to oxidize the copper.

Color Differences

It is important to note that clear lacquers will show, and sometimes enhance, subtle differences in the underlying metal. These differences often are not visible in the underlying metal, but once coated they can become apparent. When the copper alloy is polished and expected to appear as a natural copper or brass color, clear coatings can show minor differences in the metal, particularly when the metal has had the opportunity to develop a very thin oxide layer. Statuary finishes, particularly the light-to-medium levels, can also show this effect. Often the visibility of these differences is dependent on angle of view as you look through the clear layer at the reflection of the underlying metal. This is a natural phenomenon and is due to the thin layer of clear lacquer creating minor interference effects. Usually it is assumed that this is the lacquer itself, the rate of cure, or differences in thickness, but in reality the lacquer is enhancing the reflection of the underlying surface. Figure 4.13 shows two examples of copper alloy tonal differences under clear films. The image at

FIGURE 4.13 *Bottom*: Differences in metal color tone below the clear film when viewed at an angle in two surfaces. *Top*: A copper wall surface without lacquer.

the lower left shows a polished copper surface coated with Incralac, while the one on the lower right shows a surface with a statuary bronze finish. The tonal differences may be due to slightly different exposure rates prior to coating with the lacquer. The top image shows a copper wall surface without lacquer. You can see similar differences in tonal reflections in this copper surface. Uncoated, the oxide will form and the tonal differences in the base metal disappear. But in a coated surface these different reflective tones can interfere with the design expectation.

When lacquer begins to fail, it usually starts along the edges where the coating is thin due to existing surface tension as the coatings are applied and cured. A brown oxide stain forms where the coating has been breached. Incralac usually contains a chelating agent that helps bond the coating to the copper surface and counter under-film corrosion, but when a coated surface has been post-cut or post-sheared the coating will have minute cracks and will not extend over the newly cut edge.

The coatings are not self-repairable. Scratches through the coating will corrode and oxidize, showing as a darkened mark. Acrylic lacquers and wax coatings are relatively soft compared to the harder enamel coatings used on aluminum or steel.

Unprotected alloys will tarnish quickly on exposure. Whether the exposure is interior or exterior, copper and all of the brass and bronze alloys will develop a thin oxide layer or tarnish. The exception are the aluminum bronzes C61000, C61300, C61400 and nickel–silver alloys C75200 and C77000.

On exterior applications discoloration will depend on the environment. High-humidity, chloride, and polluted environments will lead to the development of surface oxidation more rapidly than other environmental exposures.

INTENT: A SURFACE APPEARANCE THAT CHANGES NATURALLY

There are many occasions when no additional protective barrier is applied to the surface of copper alloys. In this approach, the metal surface is expected to interact with the environment and develop a unique, natural, protective oxide on the surface. This could involve simply allowing the natural surface to react with pollutants in the atmosphere and grow a patina over time, or it could involve the prepatination of the surface to arrive at an intermediate appearance that will continue to react with the environment. The uniqueness of the surface lies in the reality that the surface changes the metal undergoes are "of the time"—the amount of moisture, sun, and airborne constituents the metal surface is exposed to as it develops the mineral-like surface.

This is the normal purview of copper roofs and copper walls, which when installed have the bright and shiny appearance of new copper. After a few days of exposure, they transition through an amazing array of color tones—colors created from transparent oxides growing on the surface and inducing interference effects. Depending on the humidity, after weeks or months of exposure the surface darkens to a rich deep brown tone.

Figure 4.14 shows C11000 surfaces of thin, cold rolled copper in their initial days and weeks after installation in a mild urban environment. The image at the top left shows how this thin metal

FIGURE 4.14 Newly installed thin copper sheathing.

initially appears. Surface distortion, commonly known as "oil canning," can be very apparent due to the reflectivity of the initial copper sheet. Soon, a thin oxide forms producing interference tones as light reflects off of the dual surface of the metal and the thin, semitransparent film. Eventually, as the surface oxide grows, the appearance darkens and evens out, as shown in the image at the lower left and on the right.

FLATNESS

Thin plates of all materials act as diaphragms. There is an inherent instability in thin-plate diaphragms. Sheet forms of copper alloys have an anisotropic character, which means that internal stresses are variable across the flat surface. Temperature changes can add to these internal stresses and create localized warping in and out of the flat plane as the metal wants to grow outward but is restrained by the fixed edge or the stiffened return or restricted by the interlocking clip or fold of the thin copper panel. This is what we are seeing in the top left-hand image in Figure 4.14.

For thin, flat diaphragms, thermal changes in copper alloys create dimensional changes. Copper alloys are darker and absorb infrared energy at a high rate. If the edges are confined by forming a return, the thermal changes can cause movement in the central region of the sheet such that some portions are concave while other portions are convex.

Often on thin metal roofs made of copper, distortion (the oil canning mentioned above) will be more apparent in the late afternoon sun as the surface heats up and the metal wants to expand. As the surface diaphragm grows from thermal expansion, it pushes outward and away from fixed points. When the plane is an unstable flat diaphragm, the metal expansion will be both convex and concave, even on the same surface. Because of the reflective nature of newly installed copper, this undulating surface is more apparent, as the image in Figure 4.14 shows.

In thin metals used as cladding, where the intent is for the edges to be constrained in order to provide stability against loading, the stiffness and strength of the material is less a constraint structurally than it is a constraint on the ability of the metal to move under thermal load. Usually for thin cladding material, such as roofing and flat shingle panels, a rigid backing is necessary to restrict wind loading and other pressures. Thin cladding is not capable of resisting excessive inward or positive load conditions and must have a rigid support to carry such loading.

On the other hand, thin cladding—and even more so the points of connection with the structure that are typically along the perimeter—must be capable of transferring the outward or negative load back to the structure as well as accommodating the changes in geometry created by the thermal changes.

For thicker plates, the condition is similar, except thicker plates must be capable of transferring loads through connecting points without the aid of a solid substrate. Thus, both a negative and positive load must be accommodated by the metal along with the geometry change from thermal expansion and contraction.

Figure 4.15 shows a depiction of a flat-seam, thin copper alloy fixed via clips to a rigid substrate. Thermal movement is always away from the points of fixity. It is critical that a design establish these

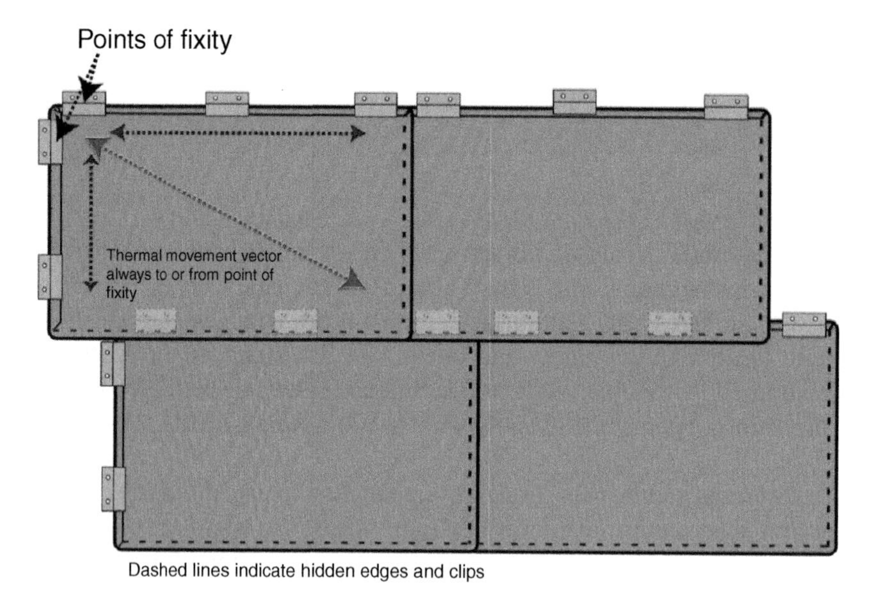

FIGURE 4.15 Flat-seam thin copper panels and expected direction of geometry changes due to temperature changes.

TABLE 4.2 Coefficient of linear expansion of various copper alloys.

Alloy	Common name	Coefficient of thermal expansion ($\times 10^{-6}$/°C)
C11000	Copper	9.8
C22000	Commercial Bronze	10.4
C28000	Muntz Metal	11.6
C61000	Aluminum bronze	9.0
C65500	Silicon bronze	10.0

points with the understanding that the geometric changes in the metal due to temperature changes is always to or from these points.

The amount of movement, regardless of thickness, is dependent on the distance or length of the metal from the fixing point, temperature differential expected, and the coefficient of thermal expansion of the metal (Table 4.2).

To determine how much thermal expansion to expect or to design for, use the following formula with the table of coefficient of thermal expansions for similar alloy types:

$$\Delta L = L_i \times \partial \left(t_f - t_i\right)$$

ΔL = Change in length expected
L_i = Initial length of part
∂ = Coefficient of thermal expansion
t_f = Maximum design temperature
t_i = Initial design temperature

As an example:
The length of the panel is 3050 mm.
The metal is copper, with a coefficient of thermal expansion of 0.0000098.
The metal is installed when the ambient temperature is 10 °C.
The highest temperature[3] the metal is expected to experience is 50 °C.

[3]Copper absorbs infrared radiation and can get significantly warmer than the ambient temperatures. It is recommended to increase the maximum ambient temperature by at least 40%.

The lowest temperature the metal is expected to experience is $-30\,^{\circ}\mathrm{C}$.

L_i = initial length of part = 3050 mm
∂ = coefficient of thermal expansion = 0.0000098
t_f = maximum design temperature = $50\,^{\circ}\mathrm{C}$
t_i = initial design temperature = $10\,^{\circ}\mathrm{C}$

Maximum thermal expansion to expect:

$$\Delta L = 3050 \times 0.0000098\,(50 - 10)$$

$$\Delta L = 1.2\,\mathrm{mm}$$

Maximum thermal contraction to expect:

$$\Delta L = 3050 \times 0.000098\,[-30 - 10]$$

$$\Delta L = 1.2\,\mathrm{mm}$$

Therefore, the design should allow for the expansion movement of the panel in the long direction to be as much as 1.2 mm and the contraction to be as much as 1.2 mm.

It is important to design for both the geometric elongation of the metal due to heat absorption and the geometric shrinkage as it cools. All materials experience this, but for metals, not designing for thermal expansion and contraction can lead to esthetic issues of oil canning and surface waving, or in the worst case, premature failure of the connections as cyclical stresses are introduced.

There are no standards for flatness. It is more an abstract. For copper and copper alloys, the initial distortions can be overcome as the metal finds its equilibrium. It will move and shift until it reaches a state where geometry changes can occur unconfined by the initial friction of the assemblies. That is assuming the design allows for that. In addition, naturally aging surfaces lose their reflectiveness as they weather, which will reduce the initial concern over oil canning.

Copper alloy sheet or plate should be leveled to remove internal stresses that have developed and accumulated during strain hardening and temper rolling. This will result in a distribution of the stresses across the sheet or plate in such a way that the stresses are negated. Doing this will provide a very even and flat sheet of metal. There are leveling methods used for copper alloys. Referred to as stretcher leveling or roller leveling, the sheet or plate is passed through a series of upper and lower small rollers that alternate moving the metal up and down. The rolls can be adjusted to subject the sheet to more deformation than other regions in the sheet in the case of differential stress. This removes the residual stress induced from coiling and cold rolling reduction process on the metal. This stress, if not removed, can show as curvature along the length of the sheet or laterally across the sheet. For heavy plate, the metal is often stretched at the Mill to flatten the plate. The result is a smooth, flat surface with minimal internal stress.

Extrusions and other linear sections, such as bar and tube, are stretched and flattened by being passed through a set of rolls or by tugging on the ends. This induces slight elongation into the part and removes differential stress that can cause shaping and curvature.

For flat surfaces, there is no substitute for thickness. Stiffeners applied to the reverse side, adhered backing, and deformation ribs pressed or rolled into the sheet help to a point, but thickness is important for flatness. Fusion studs used to attach stiffeners onto the reverse side of a thin sheet or specially formulated adhesive bonds are common means of adding section and stiffness to thin copper plates. Caution is needed to insure that there is no visible telegraphing to the face side. Again, thickness plays an important role in the stiffeners not translating to the visible surface. For the best fusion operations using these stiffening techniques, the metal's minimum thickness should be 2 mm. If the metal is thinner than this, the stiffeners will become apparent on the face surface. It takes skill and quality-assurance procedures to achieve repeatable results.

The stiffeners used on copper and copper alloys should be compatible from a galvanic-corrosion perspective. The reverse side of the copper alloy can still condensate in exterior applications, and this condensation can accumulate on the aluminum and create a galvanic corrosion situation. Normally you would simply coat the cathodic material at the interface, but in the case of copper alloys, moisture from condensation would pick up and redeposit the copper ions on the surface of the stiffeners, creating small, but numerous, corrosion cells. Otherwise, you would need to paint the entire reverse side of the copper. It is recommended to use brass or stainless steel as the stiffener material. Aluminum or galvanized steel should not be used.

The spacing between stiffeners or edges required to eliminate waviness in the face surface is shown in Figure 4.16. The vertical axis corresponds to the distance between stiffeners or the width

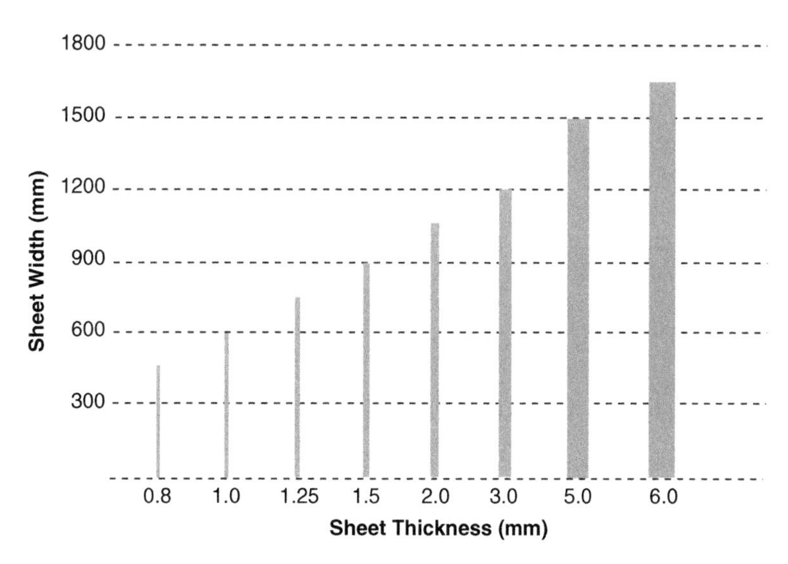

FIGURE 4.16 Sheet width-to-thickness relationship to ensure flatness.

of a panel and the horizontal axis shows the thickness corresponding to this spacing. You will need to engineer the connections with strength and stress concentrations in mind. Fusion stud welds weaken the copper alloy at the point of fusion and large flat surfaces need to be engineered to distribute the loads.

TEXTURING THE SURFACE

Adding texture to stiffen thin metal surfaces is a common way to improve flatness. Embossing, in which a metal sheet or coil is passed through rolls to impart a light texture, is one way to add texture. Figure 4.17 shows an embossed copper sheet with a light statuary finish applied. The left-hand image shows the surface embossed outward, while the sheet in the right-hand image is embossed inward.

Copper and copper alloys receive this finish well. The operation of embossing creates localized deformations to the surface, which add stiffness and allows thinner sheets to be utilized, thus reducing the cost per unit area. There are also other ways to emboss copper surfaces, which were discussed in Chapter 3.

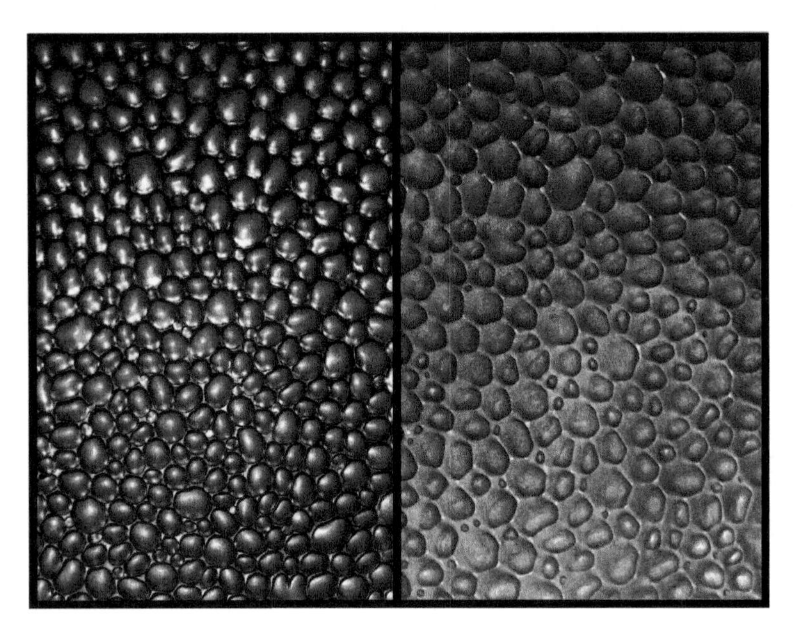

FIGURE 4.17 Examples of embossed copper sheet.

INITIAL OXIDATION ON COPPER ALLOYS

As copper and copper alloys weather, they initially develop an oxide on the surface. Depending on the humidity and the pH of the moisture, copper alloys will develop one of two forms of copper oxide—cupric oxide (CuO) or cuprous oxide (Cu_2O). In relatively low humidity and at a neutral pH, cuprous oxide normally forms on the surface. This oxide form is thin and often dark reddish in color. It can take on hues of yellow and even purple, though, as the clear film leads to light-interference behavior. In most installations, this is the expected result in the first few weeks of exposure.

The other form, cupric oxide, is dark black in color and can show as black streaks and blotches on the surface. This sometimes startling occurrence will take several weeks of weathering before the cupric oxide is converted to cuprous oxide. The black color gives way to a deep color characteristic of cuprous oxide and the surface actually lightens. The images in Figure 4.18 show the first panels set in the courtyard of the de Young museum in Golden Gate Park. The panels were installed in the winter of 2003 and early 2004. Fog from the bay and rain are common occurrences in that environment. The pH of the captured rain was acidic, measured at a pH of 4. In a matter of days, the surface appeared as if someone spray painted the surface with black paint.

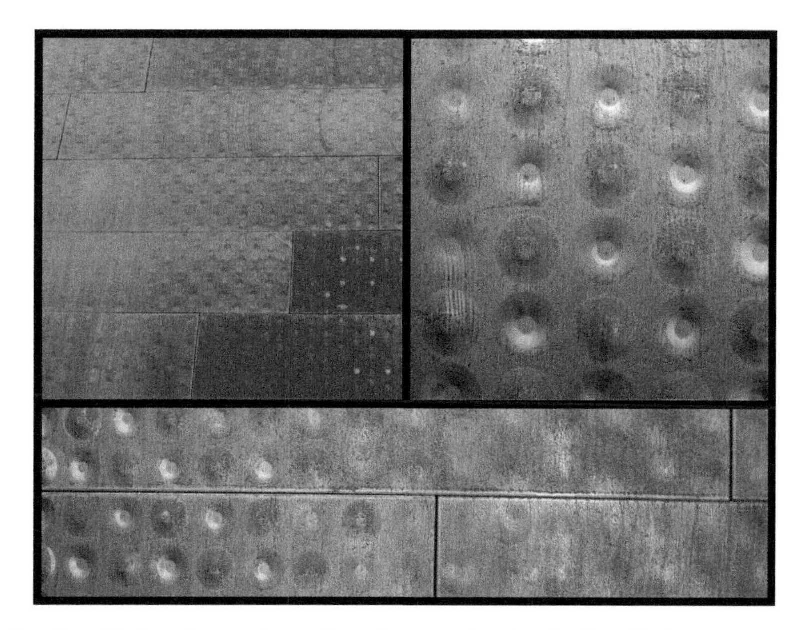

FIGURE 4.18 Cupric oxide forming on the surface of copper shortly after installation.

FIGURE 4.19 Cupric oxide changing with time of exposure.

The panels had been manufactured and coated with a clear protective film. The film was removed shortly after installation and a bright, new, unoxidized copper surface was exposed to the salt fog that came in from the bay each morning. Figure 4.19 shows changes in the surface color over time. The top image shows how the wall looked in the first weeks of exposure, while the middle image shows it after two more months of exposure. The color may have been a major concern for those seeing the wall for the first time but after a few months, as exposure to the environment continued, the cupric oxide changed to cuprous oxide and the contrast in colors was drastically reduced. The bottom image shows the wall after a few more weeks of exposure, as the contrast diminished and the color changed. Over time, the oxide was absorbing moisture and thickening while also changing to a more consistent layer of cuprous oxide on the surface.

This is not a common occurrence, but it still can develop if the right conditions are met. To hinder this phenomenon, consider exposing the copper to the air without protective coating. Allow the metal to be exposed to the air and allow an oxide to slowly develop, just enough to provide a slight tarnish of cuprous oxide. This should be done in a controlled, dry environment, such as a warehouse. Install the panels after they have been exposed for a few days and this darkening will not occur. The more stable and common transition form, cuprous oxide, will develop. The copper darkens slightly due to the cuprous oxide, but this will inhibit the development of the blackened cupric oxide. Cuprous and cupric oxides can form on any of the copper alloys, both wrought and cast. There are several factors in play that determine which of the oxides will present themselves as initial exposures happen. One is that the surface has been newly exposed to the atmosphere and another is the characteristics of the initial atmosphere experienced by the copper alloy surface. Acidic atmospheres tend to push the darker cupric oxide development.

IN SITU PATINATION

Attempts have been made to patinate a copper or copper alloy surface after it has been installed on a structure. This approach was taken in the 1960s and 1970s and based on several formulas and approaches. One such formula was used under the direction of Frank Lloyd Wright on the Price Tower in Bartlesville, Oklahoma (see Figure 3.35).

Most in situ patination processes simply do not work well, if at all. The variables that must be controlled are multiplied when attempting to pursue patination in the field. It may work with small, isolated items—such as sculptures and entryways—but large wall and roof surfaces can be a disaster.

Figure 4.20 shows two different roofs where the patina is not adhering to the copper surface. Both of these surfaces are in the Miami area. If these patinas were induced using a copper sulfide solution, the heavy chloride environment coupled with the high humidity could be a reason for the release. A chloride patina solution might have worked more effectively in this humid seaside environment. The same goes for using chloride-based patina solutions in an urban industrial environment. This criterion is only a minor constraint in the effort to arrive at an in situ

FIGURE 4.20 Failure of in situ patination.

patination of copper. There are many more challenges to achieving in situ patination on a large exterior surface.

The constraints of in situ finishing of copper alloys:

- Arriving at a clean surface
- Application of the patina solution
- Controlling humidity and temperature
- Health and safety
- Protecting the environment

For patinas to bind to the underlying metal, the surface must be very clean and free of oils and surface oxides. Often surfaces are abraded to expose the underlying metal and to give a level of roughness (or "tooth"), allowing more surface area to react. Abrasion of large in situ operations might involve blasting with sand, walnut shells, or other abrasive media. This would require protecting the surrounding areas and collecting the media. The difficulty with this on large surface areas is that one wants to blast the entire area before beginning the application. When this approach is taken, the copper alloy surface has an opportunity to develop an initial oxide at variable rates across the surface. This oxide will hamper the patination as it is applied to the surface. Some of the patina solution will be reacting with a fresh copper surface while at other areas, the solution will be reacting with a surface that has experienced oxidization. Essentially, the oxidized surface will resist the chemical reactions necessary for the patina solution to take hold. For in situ patination to work on large surfaces, you have to proceed element by element.

Applying patina solutions over large surfaces can be a daunting task. Sponging, rolling, spraying, and brushing are common approaches, but controlling the amount of solution and protecting the health, safety, and environment of the workers and surroundings is critical. Even as you are accomplishing these first steps, controlling and predicting Mother Nature are limiting parameters. The patina or statuary solution must be able to chemically react with the base metal. Diffusion must occur between the base metal and the solution. If too warm, the solution will dry out and the process won't occur. You may get an initial appearance, but it will wash off in the first rain. If the ambient temperature is too cold, the chemical reaction may be too slow. Humidity differences will also have an effect on the results.

PREPATINATION

The issues related to in situ finishing do not necessarily apply to prepatinated sheets, whose patina is developed through the diffusion of copper ions onto the patinated surface in a controlled environment. Using prepatinated copper or pre–statuary finished copper alloys rather than attempting in situ finishing is highly recommended. Use only companies that have demonstrated good environmental stewardship and exercise safe working practices. When prepatina solutions are applied, they

FIGURE 4.21 *Left*: Prepatina on interior walls at the University of Toronto. *Right*: Patinated copper door cladding on a private home.

are applied element by element. They are not paint solutions, so controlling color from one element to the other is only possible within a set criterion. A chemical reaction of the sort needed to create a patina takes time. During this time period, the temperature and humidity must be held relatively constant for several days. Sometimes the reaction appears to take place but the diffusion of copper into the patina has not been fulfilled and the patina is prone to flaking off. The left-hand image in Figure 4.21 shows an interior wall created from elements prepatinated with a chloride-based patina solution to arrive at a bluish tone. The door cladding in the right-hand image was created for a private residence using sulfur-based chemistry.

Even controlling for these variables, the color range can still be apparent. This should add to the beauty of the surface, not take it away. Patinas are like the surfaces of natural materials and not organic paints. The color and tone are variable and should be expected to occur. It should be

(*continued*)

(continued)

subtle, but color and tonal differences between elements is a natural development. It is important to understand this. Produce samples and view them at different times. Come to an understanding on the end effects beforehand to eliminate costly frustration later.

THE EFFECT OF SEALANTS

Sealants in the proximity of copper alloys will alter the development of their natural oxide. Figure 4.22 shows areas of a copper surface below glazing sealant joints. As the sealant catalyzes, oils are released that have an effect on the surface oxide. Sometimes streaks of lighter color form as the sealants decay.

This stain is indelible. It prevents oxidation from forming. To eliminate the potential for occurrence, appropriate design and detail work are necessary. Adding a drip edge, moving a sealant back behind a cover, or using sealants developed to eliminate the effect are all options for avoiding this problem.

When designing the copper–sealant interface, consider a small drip or vertical deflector to control where the oils from the sealant will drain. These will alter the way the copper alloy in the

FIGURE 4.22 Sealant streaking below glazing.

proximity to the sealant will weather. Once this begins, it is very difficult to correct. It will be necessary to remove it in order for the copper surface to age in a manner similar to the surrounding metal surface.

THE CAST SURFACE

Cast sculptures made from copper alloys can be found in every region of the world and from every time since the beginning of civilization. Techniques have been improved, but the overall process is very similar to what was done centuries ago. The foundries involved with the production of cast bronzes are experts at their trade. The metal used and the molding materials have been refined to achieve consistency in the process. There are, however, a few mishaps that can occur. When visually inspecting the bronze cast surface, flaws can be seen by the discerning eye. These flaws are due to the molding and casting process, the welding of sections, and the patina application. The casting process is an art in itself and requires a deep understanding of both the metal being cast, the volume of the final form, and the quality and nature of the mold material being used to create the form. The flaws that make it past the foundry's quality-assurance process are few, but absolute perfection is unattainable.

All large sculpture is assembled from smaller cast sections. In the beginning the artist makes a clay or wax model. This is cut into parts and fitted with clay plugs to be used as vents and pour spouts. Their location is not arbitrary. The location of seams is critical for finishing out the weld when the smaller cast sections are reassembled. Once these sections are prepared and cast, they are reassembled into the original large form. The joints are welded. For silicon–bronze cast alloys, the welding material consumed is a similar silicon–bronze metal. Usually a MIG (metal inert gas) welder set on DC current with silicon–bronze wire feed is used. TIG (tungsten inert gas) is also used with welding rod of makeup similar to the silicon–bronze sculpture. Match the wire alloy with the cast alloy.

Once the weld is achieved, the foundry will grind the weld down and blend it across the surface. If the seam is placed in a highly detailed portion of the sculpture, the grinding and finishing of the weld can be very difficult. This is a flaw that will be visible even after the patina is applied. The surface geometry can appear different and this will take away from the appearance of the sculpture. There is nothing that can be done to improve this appearance.

The welds will also weather differently from the rest of the casting. This is due to metallurgical differences between the weld area and the balance of the piece. Chapter 6 describes the welding of copper alloys, and Figure 6.23 shows some of the visible flaws that can manifest on the surface when the welds age at different rates and with different chemistry.

ARRIVING AT THE BEST POSSIBLE OUTCOME

For copper and copper alloys, the best possible outcome needs to be established early in the life of the metal. Whether you coat it with a clear lacquer or wax or allow it to age naturally, begin with

FIGURE 4.23 Marks from a marking pen on a metal after 25 years of exposure.

a clean surface. Fingerprints, oils, and marks from marking pens should all be removed from the surface of the copper alloy (Figure 4.23).

Later chapters will discuss the importance of ensuring that the surface of copper alloys are free from moisture and fingerprints before coating with a clear coating; otherwise, darkening under the film will occur. So in the beginning provide a clean surface on the copper alloy.

Begin with a clean surface.

- Allow the copper oxide to grow naturally and unimpeded.
- Prevent under-film corrosion on coated copper alloys.

Over the First Few Months . . .

For unprotected copper alloys, expect the copper to take on various colors in the first few months, as described earlier. These are interference colors and indicate a clear cuprous oxide is forming on the surface. On occasion, particularly in fog-laden environments near the sea or where acidic conditions occur, one could expect to see the formation of cupric oxide. It is not common, but it can develop. It will eventually absorb water and oxygen and form a thick layer of copper hydroxide mixed with cuprous oxide, and the dark color will disappear—still in the first few months.

If the surfaces are handled, expect to see fingerprinting and light tarnish developing, making the appearance of the fingerprints more distinct. A light tarnish will form on the surface of copper alloys, in particular the brasses. Depending on humidity levels, the tarnish will begin to form in a matter of weeks in high-humidity areas and after a longer period in dry regions (Figure 4.24).

FIGURE 4.24 Copper roof surface. *Top*: Initial installation. *Bottom*: After two years.

If the surface is polished with a light wax or metal polish the tarnish will be prevented, but be careful not to allow clumps of metal polish cleaners to remain on the surface or in seams.

Coated surfaces, both those that are prepatinated and those that are lacquered and waxed, should appear as installed over a period of the first several months to the first two years or longer. The conditions that will change the surface are salt compounds, such as those from a seaside exposure or deicing salts.

Over the Next 10 Years ...

Predicting anything that takes place beyond a short period of time is always based on conjecture, but in uncoated copper surfaces, the color tone should even out and darken as the moisture from the atmosphere and the oxygen are absorbed and the copper ions diffuse outward to begin the formation of patina. The process of developing an uncoated surface is similar to the process of raising a child: treat them well in the beginning, allow them to react to their surroundings, don't get excited about the blemishes and pimples on the surface, and they will eventually arrive at a perfectly sound surface—one unique to its time and place on the earth (Figure 4.25).

The coated surfaces should be monitored to ensure that the coating is sound and that any scratches are coated. If the surface is waxed, the wax will need to be removed in the first three to five years and a new coating of wax applied. If not, the wax will crack and shrink, some will weather

FIGURE 4.25 The surface of the National Underground Railroad Freedom Center designed by BOORA Architects after 10 years exposure.

off, and the underlying metal will be exposed to the atmosphere. When this occurs, it will darken in those spots.

To maintain the copper alloy surface, whether coated or naturally exposed, an occasional cleaning with fresh water is advised. On large surfaces such as roofs and facades, yearly inspection is advised. Rinse down, potentially with a mild detergent if necessary, to remove detrimental materials that may be decaying and reacting with the copper surface. This would include areas where steel, silicone, paper, or wood materials are in proximity and allowed to drain over the copper or copper alloy surface.

Over the Next 30 Years ...

If the surfaces are clean and free of deteriorating materials in the early years, the exposed copper should be developing a 300-plus year mineral surface. If the surface is prepatinated, expect a slight darkening as the coating thickens and forms into a rich mineral surface.

Lacquer-coated surfaces should be inspected in the first 15 years and subsequently every 5 years. Keeping these exterior surfaces clean is recommended. If cleaning is performed correctly using oxide-inhibiting lacquers, at worst they should show signs of darkening along the edges. If you are able to address this early, clean the edge with a solvent wipe and add apply more coating along this area. This will extend the life of the coating. If the area is not too large, consider a coating of hard wax, as described in Chapter 8.

Inspect the surfaces and look for loose seams or spots that could be signs of bronze disease in the tin bronzes or dezincification in the brasses. Address these when they appear to prevent a worsening of the damage. Look for areas where steel or iron assemblies are deteriorating and rusting on the copper alloy surface. These stains need to be stopped or it will turn into a permanent condition on the surface.

I know of no materials used in our environment that never need some level of maintenance. Copper roofing and walls are very close to that material. There are innumerable copper surfaces that have performed as designed for decades, even centuries, with nothing more than the cleansing from natural rains. As they age and weather, they take on their environment and achieve a beautiful, natural appearance. If prepatinated, the further enhancement of the surface should continue as if it were produced by nature.

The lacquered and the waxed surfaces will need to be maintained. It is the life of these coatings that is the determining factor in maintenance rather than the metal itself.

Wax, at best, will give from three to five years before it begins to decay, while the best lacquers can be expected to provide protection three to four times longer than a wax protective coating.

CHAPTER 5

Designing with the Available Forms

Patina is the value that age puts on an object.

John Yemma

A BRIEF HISTORY

As the oldest metal known to humans, copper has led the way for all other forms of metals over the years. It most likely was first worked by cold forging, and by this method was beaten into tools and decorative shapes or melted and cast between two rocks to create an interesting shape. Such crude wrought forms of copper began mankind's early path in the use of metals.

Ötzi, the Copper Age man whose remains were found in the Italian Alps, lived 5500 years ago. One valuable tool in his possession was a copper axe. The copper had been mined in the Tuscany region. The axe blade found with Ötzi was made from cast copper that had been honed and shaped.

For centuries copper, and then copper alloys, has been worked by mankind. The market for elaborate bronze castings established trade and industry in several zones in early civilizations from the eastern Mediterranean outward to mines in Germany and Cornwall in the west and north and eastward to Persia and India (Figure 5.1).

The casting of copper and copper alloys proliferated in antiquity. Metalworkers were held in high regard in ancient times, and often their skill and craft were secrets held within a family. In the Boston Museum of Fine Arts, you can see a stone plaque that Ramses III dedicated to Akmose,

FIGURE 5.1 *Left*: Ancient formed spear; *top right*: ewer; *bottom right*: helmet made of copper.

his chief metalworker, and his family. It is from ancient Egypt and dates to around 1450 BC.[1] The plaque reads:

> I sent forth my messengers to the country of Atika, to the great copper mines which are in this place. Their galleys carried them; others on the land-journey were upon their asses. It has not been heard before, since kings reign. Their mines were found abounding in copper; it was loaded by ten-thousands into their galleys. They were sent forward to Egypt and arrived safely. It was carried and made into a heap under the balcony, in many bars of copper, being of the color of gold of three times.

Casting copper alloys was prolific in the days of antiquity. Once ore was uncovered and mining operations and refinement were better understood, the production of copper alloy artifacts flourished. Mines in the Tarshish region in eastern Anatolia are estimated to have delivered as much as

[1]John Delmonte, Origins of Materials and Processes (Technomic Publishing Company, 1985), 209.

TABLE 5.1 Comparison of alloying constituents of Muntz Metal with Architectural Bronze.

Alloy	Copper (%)	Zinc (%)	Lead (%)
C28000	60	40	0
C38500	57	40	3

20,000,000 tons of copper ore by the time of the Romans. The metal was destined for the civilizations around the Mediterranean, and those that worked with the metal developed techniques of manufacture. Wrought copper products such as sheet copper were an early development. Sheet bronze and brass required more finesse because of their higher strength, but copper could be hammered flat into thin plates. The statue of Pepi I of Egypt was created from copper plates over wood. This was created in 2300 BCE and incorporated flattened copper rivets to hold the plates together. The methods of cold forming and hammering copper into molds were well understood by ancient mankind.

Copper was one of the first metals to be alloyed with other metals to develop sought-after attributes. Hardness and color were most likely the most interesting and beneficial attributes derived from alloying copper. The increase in hardness when tin or zinc were mixed into the molten metal had a direct influence on the usefulness of the metal. Color was another important attribute of alloying. The ability to achieve a golden color by mixing certain zinc minerals into molten copper definitely increased the interest of the metal.

Today, as in antiquity, many of the copper alloys are created to achieve the best attributes for the particular form. For example, certain additions to existing copper alloys can facilitate extrusion processes. Alloy C38500, Architectural Bronze, is very close in alloying components to that of alloy C28000, Muntz Metal (Table 5.1). The difference is 3% lead is added to facilitate the extruding process. Alloy C38500 only is available in extruded form.

The added lead enables the hard alloy to plastically deform in an extrusion press. Small amounts of lead added to copper alloys benefit many shaping processes. Other alloys can be extruded. Most have minute amounts of lead.

Attributes of Copper Alloys

- Variety of colors
- Ductility
- Corrosion resistance
- Extrudable
- Receptivity to patination
- Machinability
- Density
- Antimicrobial

- Cast ability
- Recyclability
- Electrical conductivity
- Receptivity to electroplating

WROUGHT FORMS

The wrought forms of copper and alloys of copper are similar to those of other metals. Wrought forms are those forms that have been shaped by applied force and that have a noticeable grain direction that runs parallel to the direction of the force. Plates, sheets, and strip are long ribbons or blocks of wrought metal. In contrast, the grains of the metal are aligned in cast forms, as the metal cools in a mold. No force is applied to cast forms to align and stretch these grains as there is to wrought forms, and the grains of cast forms are larger and less refined than those of wrought forms.

The wrought forms of copper alloys are:

- Plate
- Sheet
- Strip
- Coil
- Bar
- Tube
- Pipe
- Wire
- Extrusion

Not all the copper alloys are available in all forms. As shown for the various alloys listed in Chapter 2, some alloys are available in certain forms and some are not. This is due in part to industrial demand, which drives the manufacturing process at the Mill.

Additionally, there are dimensional limitations set by various constraints of the manufacturing process. There are maximum width limitations for all copper alloys, which are generally lower than those for other metals.

Mills[2] that produce the various forms of copper alloys are not the same Mills that produce steel or aluminum. Copper alloy Mills are facilities dedicated to the production of copper. One of the main production processes is the rolling and reducing of copper alloys to thin plates and sheets.

---

[2]Here the term "Mills" refer to the large industrial producers of copper alloys. They take the raw material and scrap, melt it down, and cast large blocks of copper alloys destined for further shaping in a rolling Mill, bar Mill, wire Mill, or other production facility.

Strip, Plate, Sheet, Coils, and Foil

Plate, sheet, coils, strip, and foil are considered flat rolled products. Each of these forms starts as a slab of thick alloy cast from a heat. A heat can be several tens of thousands of kilograms divided into blocks of copper alloy of rectangular or circular cross-section. The alloy constituent components are carefully controlled as each heat is produced from a single alloying mix. The grain structure and growth are examined to verify proper solidification. Usually a slice is removed, and the metallurgical properties are examined.

Figure 5.2 shows a typical cross-section of a copper casting. The large grains are exposed by etching the surface. They radiate out from the center. The outer edge cools more quickly and so the grains are smaller than the more-slowly cooling center.

There are two main ways that plate, sheet, and strip are produced: the vertical direct-chill cast method and the horizontal continuous cast method. The vertical cast method produces large slabs of rectangular or circular cross-section. The rectangular cross-section blocks are destined for the rolling Mill, where they will be heated, hot rolled, and reduced in thickness as they are increased in length, then coiled into rolls (Figures 5.3 and 5.4).

The horizontal continuous cast method produces a thinner cross-section in long lengths, which are directly rolled into coils. As the metal thins it hardens and cools. This work hardens the metal and heating is necessary. Heating anneals the copper alloy and achieves grain refinement. More consistency and smaller grains are achieved. Temperature control is critical at this juncture for copper alloys. There is a very tight temperature range for each alloy.

For the various copper alloys, the Mill produces metal based on need. For large architectural copper alloy projects, there are Mills that will work with the design to refine the metal and tailor the

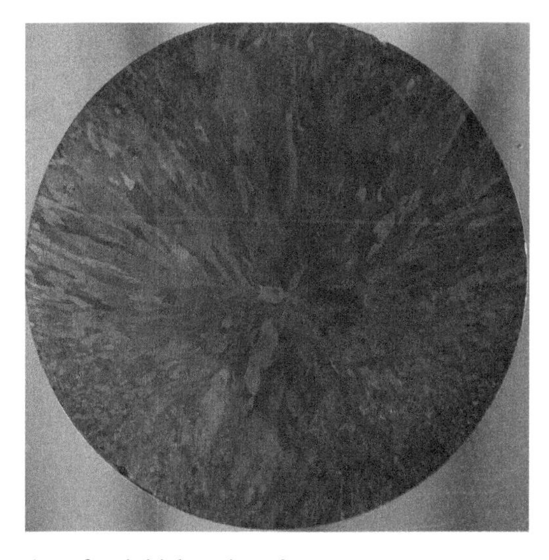

FIGURE 5.2 Etched cross-section of an initial casting of copper.

FIGURE 5.3 Large slab of copper at the rolling Mill.

FIGURE 5.4 Large copper coils.
Source: Photo courtesy of KME.

alloy and its mechanical properties to specific needs. Smaller projects, and art work in general, are subject to supply constraints. There are some facilities that stock specific alloys in specific sizes for various clients and industries.

Accommodating a request for a specific alloy and mechanical property is not always possible.

Supply Constraints

- Minimum quantities
- Availability
- Stocked sizes and thicknesses; alloys and tempers
- Surface quality

Plate

Not all the copper alloys are readily available as plates. If there is little-to-no demand for a particular form, Mills will not produce the alloy in that form. There is no on-demand casting of large slabs of specific alloys. That being said, all sheet material begins in a plate form. It is the production process of making the sheet that cannot be altered easily to provide plates in small quantities.

Thick copper plates of high purity are used extensively in the electrical industry, where they are sized for a specific amperage. Alloy plates are also machined into fixtures for various industries needing strength and corrosion resistance.

Due to the relatively high price, copper alloy plates are not often used in art and architecture. Copper alloys are dense, and for a metal that is often priced by weight, the cost of thick plates can quickly engulf the project budget. Depending on market conditions, brass alloy plates can often be slightly lower in cost than copper due to the more economical zinc. Couple this with the increase in strength the brass alloys provide, and a design may work with a thinner plate (Table 5.2).

Thicknesses of 6 mm (0.24 in.) and thicker are considered to be plate. At the Mill, plates are hot rolled to thickness and often given a final cold pass to produce a smooth surface finish. The surface finish on plates is smooth and often level, but the outer layer will be covered with oxides. As the thickness increases to around 50 mm, the surface can be rough with thick oxide scale. Not all of the alloys are readily available in plate form. The thicker the plate, the less smooth the surface. Check the available thicknesses for a particular design; alternative alloys may be required to meet the design needs.

Plates wider than sheet copper alloy widths can often be obtained. The standard width of copper plate is 1250 mm (49 in.); however, you can obtain wider plates in certain alloys and from certain custom manufacturers (Table 5.3). The finish will not be as smooth because the plates will not be passed through cold rolls. Additionally, due to the higher cost, obtaining samples may be difficult, since they would have to be cut out of a large section. The weight will be significant and the cost per unit area for a sample will be considerable. Figure 5.5 shows a large, 50-millimeter-thick copper

TABLE 5.2 Approximate weights for copper sheet and plate.

Nominal thickness (mm)	Approx. weight (kg/m²)	Nominal thickness (in.)	Approx. weight (lb./sf)
1.0	9.01	0.04	1.86
1.5	13.51	0.06	2.79
2.0	18.02	0.08	3.72
2.5	22.52	0.10	4.65
3.0	26.57	0.12	5.49
4.0	35.54	0.16	7.34
5.0	44.59	0.20	9.21
6.0	54.05	0.24	11.16
9.5	84.45	0.38	17.44
13.0	112.61	0.50	23.26

Brass is approximately 6% lighter than copper depending on the density. Copper alloys are relatively dense and usually priced by weight as opposed to surface area.

TABLE 5.3 Typical widths available for plate.

Available widths	
Millimeters	**Inches**
914	36.0
1000	39.0
1250	49.0

FIGURE 5.5 Thick copper plate.

plate, degreased and blackened for the artist John Lajba. This plate was saw cut out of a copper slab approximately 1780 mm wide and finished to a deep black under the direction of the artist.

Sheet and Coils

Copper and most copper alloys are produced in sheet form. Sheet can be of many different lengths, since sheet stock come out of coils produced at the rolling Mill. Typical sheet is available in the common sizes that correspond to most manufacturing equipment and shipping skids. Sheets are readily available in what are considered standard lengths (Table 5.4).

Sheets are defined as rectangular cross-sections of thicknesses up to 4.8 mm (0.188 in.) and of widths wider than 500 mm (20 in.), cut to length, leveled, and flattened (Table 5.5).

> The limitations on sheet dimensions are defined by industry. Widths are limited by Mill cold rolling stations. Lengths are limited by shipping and packaging constraints and the equipment used to manufacture the final product. Special skidding and packaging are needed if lengths exceed 3.6 m (12 ft.). Most equipment used today for shearing and forming are limited to handling sheets that are a maximum of 3.6 m (12 ft.) wide.

The use of copper as a vertical surface material in architecture has created a large market for the metal in the last several decades (Figure 5.6). The beautiful color tones of copper and patinated

TABLE 5.4 Standard lengths of sheets.

Meters	Inches
3650	144
3050	120
2440	96

TABLE 5.5 Typical widths available for sheet stock.

Available widths	
Millimeters	Inches
914	36.0
1000	39.0
1250	49.0

FIGURE 5.6 Copper panels on the de Young museum in San Francisco.

copper have moved the metal into areas previously not considered. Flat, interlocking seam, thin copper panels follow many of the rules of standing seam roof systems. A rigid solid backing material is necessary to do the heavy work, and the copper provides a formidable barrier to the environment.

Rain screen panels are finding interest as well in the design community. These require thicker copper, often with higher yielding tempers. The metal must be sufficiently thick to handle wind loads and to not show excessive oil canning deflections on the face.

The C11000 alloy is a common architectural roofing material, and it is being used more and more frequently as a material for the walls of structures. Copper can be provided in coils to be attached to a roll forming station and shaped into standing seam or batten seam roofing. Roofing copper, as it is sometimes called, is nominally 0.56 mm (0.022 in.) in thickness. In the United States it was once commonly referred to as 16-ounce copper and was provided with a quarter-hard temper. Figure 5.7 shows three types of copper roofing: batten seam, curved standing seam, and horizontal Bermuda seam.

FIGURE 5.7 Various copper roof types. Top left: batten seam; top right: standing seam; bottom: Bermuda seam roof configuration.

Copper roofing and flashing coils are available in the C11000 alloy. Coils in the alloy C12200 are available but not typically used in roofing. The C12200 alloy, known as deoxidized high phosphorus (DHP), is ideal when brazing and welding operations are needed.

As mentioned previously, a carryover from past practices meant that descriptions such as 16-ounce copper and 24-ounce copper were once common in the United States. These terms correspond to the approximate weight per square foot of copper, which in turn corresponds to a specific thickness. Table 5.6 shows the correlation of the weight thickness referenced by these terms to nominal length thickness. Note that metals are a global industry and most metal facilities are moving to a nominal range associated with metric dimensions.

Copper and many of its alloys are available in coils. Coils can be slit to the strip widths to accommodate particular industry requirements or they can often be found in stock lengths. Note that not all copper alloys are readily available in coil. Mill requirements may subject an order to a minimum

TABLE 5.6 Correlation of weight thickness to nominal length thickness.

Old US description (oz.)	Thickness (mm)	Thickness (in.)
16	0.6	0.022
20	0.7	0.027
24	0.8	0.032
32	1.0	0.040
48	1.5	0.060

quantity and a particular width depending on the Mill's cold rolling widths. Verifying availability with the Mill is recommended.

Copper alloy C11000 is available in coils for various applications, roofing being a common one. Widths of coils range from 305 to 1220 mm (from 12 to 48 in.) and weights of the coils can vary from 22 to 44 kg and heavier depending on the equipment.

Many of the brass alloys are also available in coils. For most art and architecture applications, brass and bronze are supplied in sheets due to the surface finish requirements of the designs. Coils are used in stamping operations and continuous embossing operations where economies can be obtained.

Strip

Strip is a common term used for forms of copper and its alloys that are provided in narrow coils and produced in the same way as coils are used to produce sheet. In other words, strip is simply a narrow coil product that has been cold rolled to a specific thickness and temper. The maximum thickness of strip material is 4.8 mm (0.189 in.), the same as that defined for sheet. The maximum width of strip coil is 305 mm (12 in.). These small coils are destined for various industries that stamp or shape copper alloys into commodity products. The strip coil form facilitates production processes that are designed to receive the small-width coils.

Foil

Copper and a few copper alloys are also available in foil form.

Copper foil is used in the manufacture of printed circuit boards. There are brass foils available, the most common being alloy C26000. Brass foil is available in widths similar to strip, because essentially alloy foils are very thin strip forms.

Copper foil, however, is different. It is produced two ways: cold rolled—called RA (rolled annealed) copper foil—and annealed—ED (electrodeposited) copper foil. The difference between the two is the way in which the grains are aligned, which will have an effect on their mechanical characteristics.

TABLE 5.7 Nominal foil thicknesses and unit weights.

Thickness (mm)	Thickness (in.)	Weight (oz./sf)
0.018	0.0007	1/2
0.025	0.001	3/4
0.036	0.0014	1
0.076	0.003	2
0.127	0.005	4
0.177	0.007	5
0.254	0.010	7.5

The grains in ED copper foil run perpendicular to the foil due to the way the copper is electrodeposited. The grains grow outward as the copper comes out of solution, which yields a tougher, less ductile foil form with higher tensile strength and less elongation under load.

RA copper foil is produced similarly to strip and sheet. It has lower strength and is more ductile. The grains run the length of the foil as opposed to perpendicular to it.

In the United States copper foil is often sold using the same units as copper sheet, in ounces per square foot. This is a carryover from the past practice of defining copper sheets by unit weight per unit area (Table 5.7).

Leaf

Copper leaf and brass leaf are thin foils. The brass is often called imitation gold leaf. Copper leaf is more common and can come with striking colors induced into the surface. Figure 5.8 shows an example of some of the color tones available for copper leaf.

Copper and brass leaf comes interleaved in supporting tissue paper. Copper and brass leaf is only 0.0008 mm (0.00003 in.) thick and comes in packs of 25 thin squares that measure either 140 × 140 mm or 160 × 160 mm (5.5–6 in.). Like all metal leaf, copper leaf will tear easily and must be applied with sizing and coated with a clear seal afterwards.

There are a number of variations in color and finish available for copper leaf. Leaf with induced color patterns is sometimes referred to as "variegated" leaf, produced with reds, greens, blues, blacks, and mixtures of color are available.

Extrusion

Copper and many copper alloys can be extruded. There are two distinct forms of extrusion: hot and cold. Cold extrusion uses a process similar to forging. Most copper alloy extrusions destined

FIGURE 5.8 Copper leafing.

for art and architecture are produced through hot extrusion. The process is not as straightforward as extruding aluminum. For one thing the temperatures required are significantly higher (usually 595–995 °C, or 1100–1895 °F). Also, the surface oxidizes when the hot metal exits the die so more final preparation is needed.

There are several common alloys that make up the bulk of extrusions. Pure copper alloys and several brass alloys can be extruded. The pure copper alloys are often extruded for custom electrical applications, while the most common alloy of brass used for extruding in architecture and art is C38500. Known generally as Architectural Bronze or leaded Muntz Metal, this alloy is frequently extruded for architectural handrails. See Table 5.8 for a list of several alloys that are available in extrusion form.

Other alloys can be extruded, and some such as aluminum bronze, can be extruded but with great difficulty. Others are not available in extrusion form and may be better suited to have the desired shape milled from bar.

Extrusions can be solids or hollows depending on the design requirements. Unlike aluminum, whose maximum extrusion dimensions can be in excess of 250 mm, copper alloys require more

TABLE 5.8 Several extrudable copper alloys.

C11000	Copper
C21000	Brass
C23000	Brass
C24000	Brass
C38500	Brass
C65500	Bronze

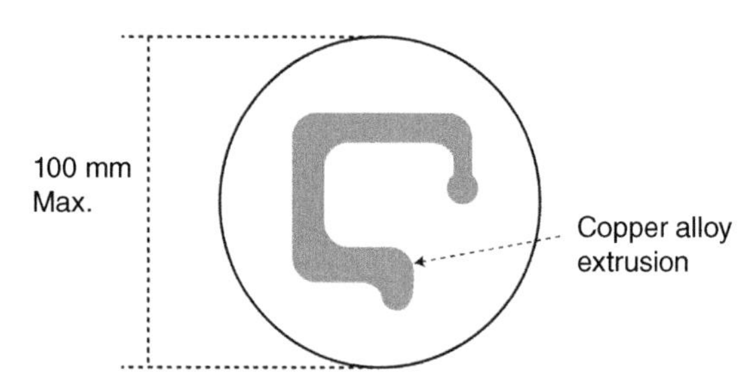

FIGURE 5.9 An extrusion must fit within a circle whose size is determined by factory capability.

pressure and more heat. The maximum size of a given cross-section must fit within a 90–100 mm circle and depends on the extrusion factory (Figure 5.9).

The Extrusion Process

All extrusion processes begin with a cylindrical billet of solid copper. The billet is composed of a select alloy of copper created to specification at the Mill source. Several of the copper alloys have small amounts of lead in them to facilitate the hot shaping of the metal as it extrudes. The billet is often cut down from a larger cylindrical casting called a "log." Billets can range in length from 660 mm (26 in.) to as large as 1830 mm (72 in.) and in diameter from 76 mm (3 in.) to as large as 838 mm (33 in.) depending on the equipment set to receive them. The larger diameters require significant tonnage in the press used to push the heated copper through the die. The maximum diameter is 100 mm (4.5 in.).

The billet of metal is heated and the equipment (the press and die) are preheated. The billet temperature is brought up just short of the melting point of the copper alloy. Pressure is applied to the end of the billet by means of a special hydraulic extrusion press. The pressure can range from 100 tons to as high as 15 000 tons depending on the equipment. This pressure pushes the hot billet up against the steel die. As the pressure builds, the semiplastic copper flows through the die opening and conforms to the shape cut into the die.

As the hot extrusion exits the die it is run out onto a set of roller tables. As it is brought out on the table it is cooled by a water quench or air wash. The extrusion may twist due to differential cooling or out-of-balance stress due to the shape. It is cut to length to fit into a stretching device. The length is stretched 1–3% to eliminate the stress and straighten the extrusion. The ends may be trimmed to a finish standard length.

Most copper extrusions are thick and compact. Thin, spindly forms will not extrude well due to the intense heat. They can collapse and bend.

Designing the Shape

The shape, also known as the profile, has several constraints that will affect whether the part can be successfully extruded, the rate of production, and the ultimate cost. Similar constraints exist with other metals; however, copper alloys are extruded at much higher temperatures.

There are three basic categories of extrusions: hollow, semihollow, and solid. A hollow form has a perimeter enclosing a void or series of voids (Figure 5.10a).

A semihollow extrusion is not totally enclosed and has what are referred to as tongues, areas of perimeter metal that extend into the void but do not totally enclose the void (Figure 5.10b).

A solid extrusion may have insets or protrusions from a central core. It could be rodlike in profile, but there is no significant void region. Structural forms such as I-beams, angles, and channels are solid extrusion forms (Figure 5.10c).

These categories of extrusion form all are affected by constraints that will determine if additional features must be added or changed in order to produce a successful extruded part. The hollow forms

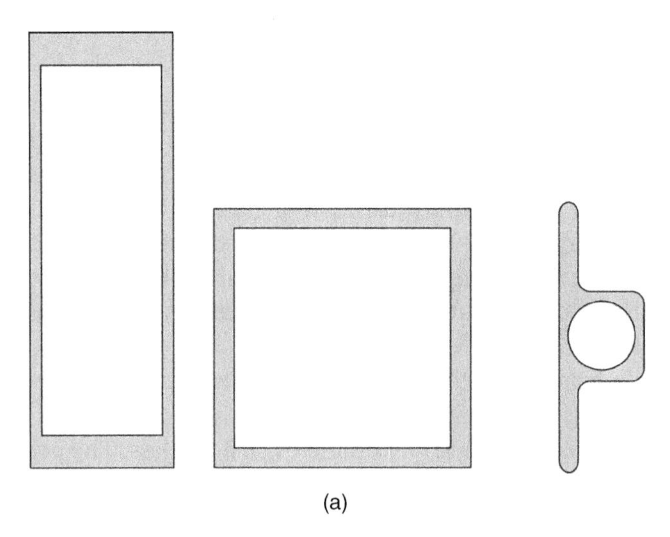

FIGURE 5.10 (a–c) Examples of hollow, semihollow, and solid extrusions.

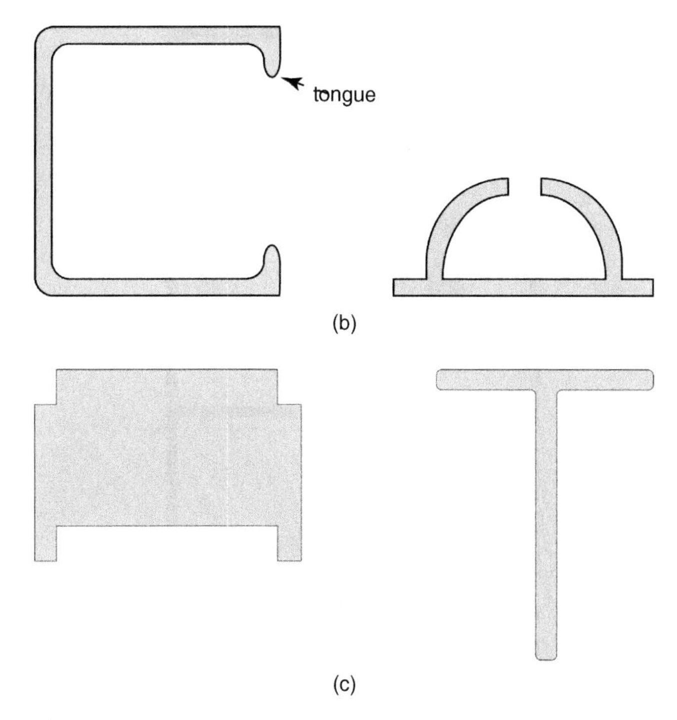

FIGURE 5.10 (*continued*)

require special dies. The way the metal will flow as it passes through the die in the process of creating the hollow form must be understood.

Cross-Section Design Criteria

The first of these process constraints is whether the profile of the overall cross-section can fit within a given circle size. This constraint is referred to as the circumscribing circle size, or CCS (Figure 5.9). For copper alloys, the CCS is limited. Working closely with the fabrication company to verify what size can be accommodated is advised. There are larger presses that can accommodate larger diameter billets and thus larger profiles, but these will require a significant increase in tonnage and will be more expensive. Additionally, larger size profiles require more care and work as they exit the die. They may twist more and may show more flaws induced from the die or from variations in the billet. Therefore, if the design is going to incorporate a large single extruded form, working with the extrusion facility and getting their input on the design is highly recommended.

If the design calls for a part that is larger than a given circle size or if the cost of using the larger die is high, consider creating two profiles and assembling the part from two separate extrusions pushed through a smaller circle size. Figure 5.11 shows a box extrusion made from two sections.

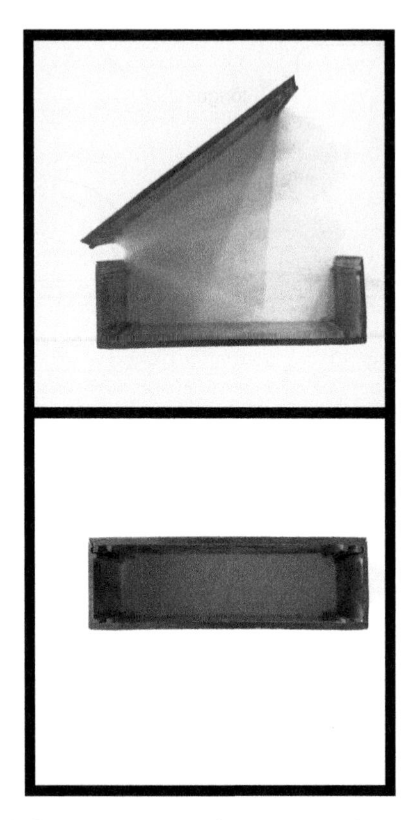

FIGURE 5.11 A box extrusion made from two sections that snap together.

Rules of Designing with Copper Alloy Extrusions

- Minimize asymmetric shapes
- Balance wall thicknesses across the cross-section
- Minimize hollows
- Round the edges
- Add grooves, offsets, and ribs to aid in flow balancing
- Use thickened walls
- High tongue ratio: the width of opening/height of protruding elements
- Achieve economy with quantity
- Review design with the extrusion company

Copper alloy extrusions are a unique design form for the artist and architect. One can incorporate strength while eliminating redundancy while at the same time adding features with specific esthetic traits. In designing with extrusion, understanding the process and the constraints are important for success.

Extrusion profile design is intuitive when you consider how a viscous fluid wants to move through an opening. When the opening is out of balance, the thicker region puts up less resistance than the thinner region and this causes imbalance in pressures. The larger area has less resistance, and this is where the metal will want to go. This behavior applies to sharp edges and corners also, which put up resistance to the flow and should be avoided.

You can facilitate the flow of metal through the die by adding grooves, webs, ribs, and rounding edges (Figure 5.12). This is not putting unnecessary metal into a design, but allowing a design to be thinner and more detailed.

Copper alloy extrusions are limited in their maximum dimension. Building up assembles with concealed joints is a way to achieve larger pieces with detail. Creating functionality with extrusions can save time and money. Adding indexing features, screw guides, and screw receivers allow for the joining of multiple extrusions into larger assemblies. Extrusions can be designed within tolerances that enable one piece to snap into another to produce a hollow shape without the higher expense of a hollow shape die (Figure 5.13).

Creating larger parts from extruded assemblies is another way of reducing cost. An elegant curtainwall cladding for the Ohio Supreme Court Building, in Columbus, Ohio involved creating a large section from a series of identical extrusions. The die was designed such that when assembled it a larger surface form would be created. Another advantage of copper alloy extrusions is that they can be oxidized in a fashion similar to other wrought forms. Polished, statuary, and patina finishes can be created on the surfaces of extruded copper alloys. The extrusions shown in Figure 5.14, used for the exterior mullions of the Ohio Supreme Court Building, all were statuary finished to a dark bronze color and then softly rubbed to induce a directional highlighted tone.

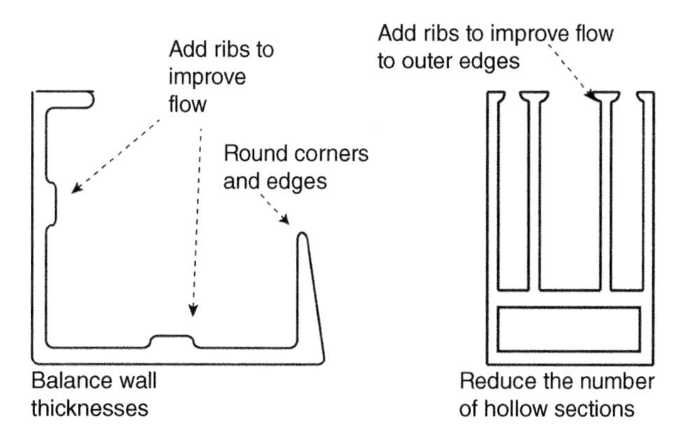

FIGURE 5.12 Improvements to extrusion design to aid in metal flow.

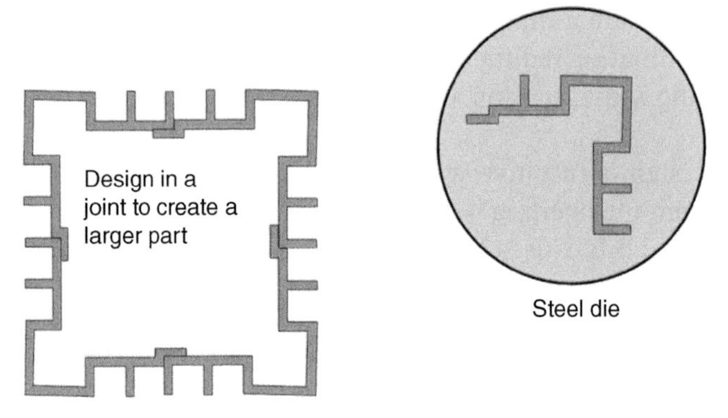

FIGURE 5.13 A larger part created from a series of smaller extrusions.

FIGURE 5.14 Extruded brass with statuary finish on the Ohio State Supreme Court building in Columbus.

FIGURE 5.15 Entertainment Building, Hong Kong.

The use of extruded copper alloys as a design option is less common because of the proliferation of aluminum extruding facilities and the low cost of aluminum, the need to coat the exposed copper alloy, fears about using dissimilar metals, and a lack of understanding of how to build and design with the material.

A statuary bronze tone on a copper alloy can enhance its elegance and beauty, producing a result that no other metal can come close to. It does require maintaining the surface and separation of less noble metals, but this is the case for other metals as well. Figure 5.15 shows a curtainwall cladding on an office tower in Hong Kong Central. This surface has been exposed for over two decades with little maintenance. The parts were made from extruded C38500, then statuary finished to a deep bronze color.

Pipe and Tube Forms

Copper alloys are available in pipe and tube. Tube forms are more common in art and architecture applications. Pipes are vessels used to transport fluids, and therefore are engineered for this purpose. Copper and brass pipe are available up to approximately 305 mm (12 in.) in diameter.

Tubes are engineered and fabricated to precise dimensions. Copper alloy tubes are available in numerous diameters and various thicknesses. Copper tubes are manufactured in three standard thicknesses up to 305 mm (12 in.) in diameter. Tubes are available in the following copper alloys:

C11000	Copper
C22000	Commercial Bronze
C23000	Red brass
C26000	Cartridge Brass
C27000	Yellow brass
C27200	Yellow brass
C27400	Yellow brass
C28000	Muntz Metal
C33000	Low-lead brass
C33200	High-lead brass
C37000	Leaded Muntz Metal
C44300	Admiralty Brass

Copper alloy tubes can be finished much the same way as other wrought copper products. Satin finishing, bead blast texturing, statuary finishing, and patination can be achieved on the surfaces of copper alloy tubes.

Bar and Rod Forms

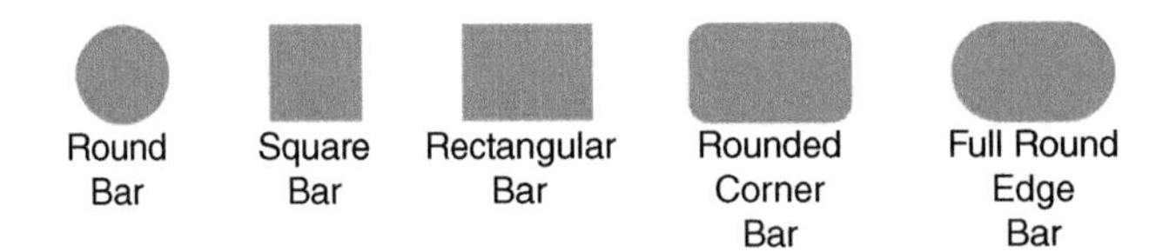

The versatility of copper alloys and the many forms these alloys come in afford the designer some creative options few other materials can match. Their malleability, strength, and coloring potential have made this the metal of choice for artists since its earliest discovery.

Bars and rods made of copper alloys can be hammered and shaped, machined, soldered, brazed, or welded together to create custom forms that still possess strength and durability.

Solid copper alloys are available in bar and rod forms from as small as 1 mm^2 or 1 mm in diameter to as large as 100 mm in diameter—or even larger for some alloys. High-purity copper alloys are used in the electrical industry and various busbar dimensions are produced in very large cross-sections designed to carry massive amounts of electrical current with reduced resistance.

"Rod" is a term applied more often to circular cross-sections. Bars can have cross-sectional geometries of any form. Bars and rods will twist when subjected to bending or torsional loading – which doesn't mean they cannot be subjected to such loads, only that they are weak in these

FIGURE 5.16 Copper bar used as handrails at the Smithsonian's National Museum of the American Indian. . Source: The handrails were designed by Ilza Jones

directions. Many a hexagonal bar or milled circular bar has been subjected to torsional loading in various applications.

Figure 5.16 shows handrails at the Smithsonian's National Museum of the American Indian made from solid rectangular copper bar. The bar was milled to an oval form, shaped by pressure and heat, welded, and then darkened to a deep brown oxide.

Bar and rod come in a number of sizes, but it is important to verify the availability of various cross-sections and stock lengths as well as tempers. The temper can be adjusted in forming operations, as cold working the shape increases the temper. Significant welding or brazing, however, will soften the temper. Heating the copper alloy to allow for severe forming operations is a common practice used when working with copper alloys in art and architecture. The images at the left of Figure 5.17 show 9.5-millimeter copper round bar shaped for the National Museum of the American Indian. The right-hand image shows a drawing of a beautiful but delicate Sidney Gordin sculpture made from silicon bronze 2-square millimeter bar.

Most of the copper alloys used in art and architecture are available in bar forms. The forms can be finished in a mirror satin finish with a directional bias, glass bead, or Mill surface. If a Mill surface is requested, be aware there may be some discoloration from packaging and minor surface marring. If a finish is required, it will be necessary to involve secondary processing facilities. The statuary finish or patina is usually applied post-fabrication.

FIGURE 5.17 Examples of different sized bars and rods used in art applications. The left is solid, round rod. The right image is very small, 2 mm square bar.

When copper alloy bar is provided with a finish it should be protected, free of all oxides, polishing compounds, scratches, and mars. It should arrive in protective packaging and be dry and free of oil. Otherwise the finish will be no different from, and potentially of lower quality than, a Mill surface.

Wire and Wire Rope Forms

Most of the copper alloys used in art and architecture are available in wire forms . Copper wire and wire rope production is centered around the needs of the electrical industry. Wire of various dimensions (or gauges, as they are referred to in the United States), braided wire rope used for shielding and electrical passage, and stranded wire cable for strength and electrical conduction are all readily available. For art and architectural purposes, the designer can employ the wire forms developed for this market or seek out a supply of special alloy wire. Special alloy configurations are not standard in the industry, thus low volumes will restrict the availability of the more custom alloys in wire rope

TABLE 5.9 Several of the copper alloys available in wire form.

Alloy	Category
C11600	Copper with silver
C14415	Copper with 0.15% tin
C14425	Copper with 0.30% tin
C23000	Brass
C26000	Brass
C27000	Brass
C27200	Brass
C51000	Phosphor bronze
C65500	Silicon bronze (MIG wire)
C75600	Nickel silver
C76400	Nickel silver

or braided wire rope forms. However, there are several alloys typically available to the designer in basic wire form (Table 5.9). Note that pure copper alloys are designed for electrical conductivity and contain trace amounts of other elements to improve strength.

Copper alloy wire rope comes in various strand configurations based on how the wire is braided (Table 5.10). Availability of the various stranded configurations is subject to manufacturer limitations and volume limits. This is a volume operation, which involves set-up costs and minimum runs. Commercially pure copper stranded wire is readily available in most configurations; configurations are mostly limited for copper alloys and keyed to certain industries, such as the jewelry or the marine industries.

Copper alloy wire rope is very flexible and strong. There are several artistic ways of working with stranded copper alloy rope. The designer of the copper handrails at the National Museum of the American Indian wanted the copper assemblies to look as if they were lashed together, to emulate rope. Copper wire rope was wrapped around the connections and then brazed to fix its position, after which it was darkened and polished to highlight the high and low points (Figure 5.18).

An artistic use of wire rope involves adding a layer of brazing alloy, such as silver solder, over the stranded wire rope. You can shape the copper wire rope into a form, then freeze the shape by adding braze to the outer surface. It takes an appearance of bark on a tree. "Branches" can be wrapped around and extend from the main trunk of this brazed tree. This weakens the rope from a structural standpoint but stiffens the assembly and produces an intriguing organic appearance.

TABLE 5.10 Various strand configurations.

Cross-section	Type	
	1 × 7	1 wire, 7 times
	1 × 19	1 wire, 19 times
	7 × 7	7 wire bundle, 7 times
	7 × 19	19 wire bundle, 7 times

FIGURE 5.18 Copper wire rope used as an artistic feature on the handrails at the National Museum of the American Indian.

Copper Alloy Mesh and Expanded Metal

The use of expanded metal as a surfacing material in art and architecture has increased in acceptance. The use of expanded metal has grown beyond strictly industrial uses to provide elegant shading and texturing to surfaces. "Expanded metal" is the term used for screenlike material that has been pierced in a grid across the entire sheet, then stretched to elongate the piercing into a regular pattern of perforations. The perforation takes the form of shell-like or elongated diamond-like openings in the metal surface. Three examples are shown in Figure 5.19.

One major benefit of this form of perforating is the use of material. No material is removed in the process, and when the metal is pierced and stretched the coverage of a given sheet is increased. For example, a sheet that is a meter wide by several meters in length, once pierced and expanded, will be wider or longer depending on how the metal is stretched.

All the sheet forms of copper and copper alloys can be expanded in this manner. Patinated sheet as well as statuary finished copper alloy sheet can be pierced and expanded.

In a diamond perforation, the openings can have a raised "lip" form that protrudes in one direction out of the plane of the sheet. This is called "standard expanded metal" or "formed expanded

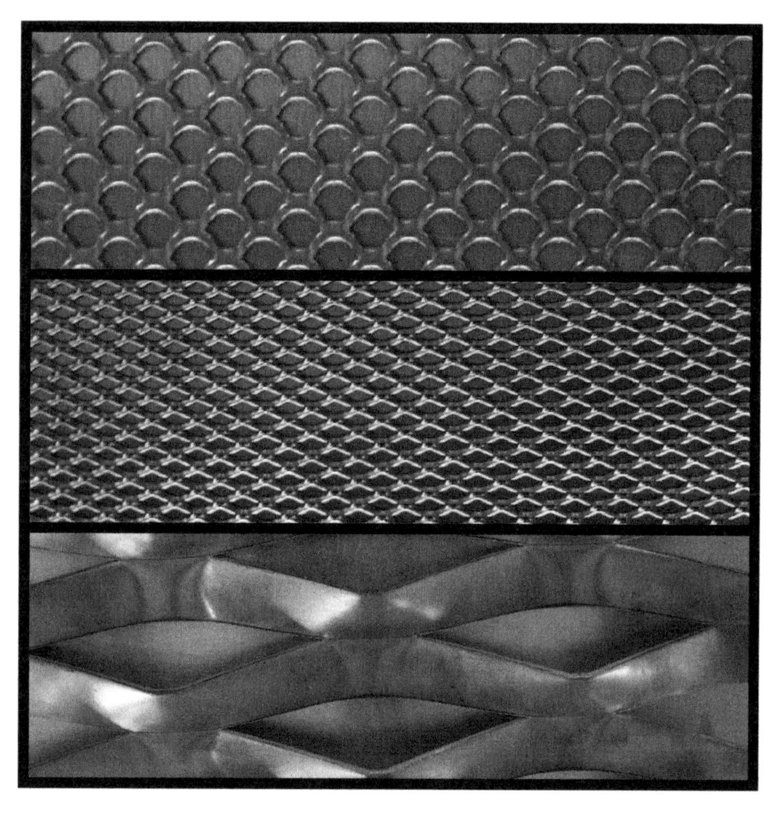

FIGURE 5.19 Three examples of expanded copper alloy.

metal." This adds stiffness and strength to the sheet of metal. It can be flattened in subsequent operations. This is called "flattened" expanded metal. The shell patterns are typically provided in the flattened form.

The expanded diamond pattern can run across the sheet or lengthwise, and can be arranged in a straight pattern grid or a staggered grid. The diamond size can come in various sizes, from small to large, with only one size per sheet. Copper expanded metal can be patinated. Figure 5.20 shows an incredible use of expanded copper with a patina. This project, designed by O. M. Architekten

FIGURE 5.20 Expanded patinated copper surface created from KME's TECU Patina Mesh. Image courtesy of KME.

BDA in Braunschweig, is located on the Innovations Campus der Wolfsburg AG, Wolfsburg, Germany.

Copper alloy mesh or screen is available in coarse, medium, and fine weaves and in various wire thicknesses. Approximate sizes are in the following ranges:

Coarse weave	2 mm spacing to 25 mm spacing
Medium weave	0.5 mm spacing to 2 mm spacing
Fine weave	0.03 mm to 0.122 mm spacing

Wire sizes range from 3 mm (0.118 in.) in diameter for the coarse mesh to 0.030 mm (0.001 in.) for the really fine meshes. Wire sizes are often given in wire gauge units. For example, gauge 10 is the equivalent of a diameter of 2.6 mm (0.102 in.), while 18 gauge wire equates to a diameter of 1 mm (0.04 in.) and 30 gauge to a diameter of 0.01 mm (0.25 in.). The larger the gauge number, the smaller the diameter. It is a logarithmic scale developed years ago by the Swedish corporation Brown & Sharpe, for which the scale is named.

Commercially pure copper is the most common copper alloy used in the production of mesh forms. Copper has good corrosion resistance and is used widely in insect screening and in Faraday cages under floors and in walls. Other alloys are available subject to volume and production limitations: essentially any of the wires can be woven into mesh material but they are not in ready production like copper mesh and screening.

Woven wire mesh is categorized by wire size and mesh count. Mesh count corresponds to the number of openings in a square centimeter or square inch.

Meshes can be shaped easily and the fine meshes can be cut with hand shears. Shaping the mesh surface creates an interesting contrast of light and dark regions corresponding to the soft creases. Figure 5.21 shows an example of fine mesh being used as a decorative light diffuser. Figure 5.22 shows three examples of woven wire meshes available in the market.

Decorative meshes that incorporate different copper alloys, flattened wires, and are even mixed with stainless steel wire are used as high-end surfacing materials for elevator cab walls and other wall treatments where the combination of decorative elements and durability is needed.

Perforated and Embossed

Copper alloy sheet is an ideal material for perforating and bumping. The metal is softer, and the lower yield requires less energy. This leads to fewer localized stresses being imparted to the surrounding metal. Because copper readily develops a visible oxide, the edges will darken and eventually patinate to be similar to the balance of the surface. The considerations involved in purchasing perforated sheets of different thicknesses and alloys are similar to those that apply to expanded copper sheet. There are many toll perforation companies that can provide different patterns of holes using different copper alloys and different thicknesses.

Copper and copper alloys can be embossed with various patterns. The beauty of this embossed treatment, however, comes into play when the metal is further oxidized. See Chapter 3 for more descriptive information on oxidation and patination of copper alloys.

FIGURE 5.21 Fine mesh light diffuser.

FIGURE 5.22 Decorative copper alloy meshes.

Custom Perforation of Copper and Copper Alloys

Copper can be custom perforated. Custom perforation involves piercing with a tool and die controlled by computer-driven equipment. Turret punch and similar equipment use less pressure to pierce an equivalent hole in copper than in steel. Additionally, there is less work hardening, so the metal remains flat as the piecing occurs. The custom pattern used on the Irving Convention Center designed by RMJM & Hillier (Figure 5.23) is an excellent example of the use of custom perforation, adding both function and beauty to the surface.

FIGURE 5.23 Custom perforation of the C11000 alloy used in the Irving Convention Center, Las Colinas, Texas. The structure was designed by RMJM & Hillier.

FIGURE 5.24 The C11000 alloy with a Dirty Penny finish, surfaced with bumps and perforations and used in an exterior (*top*) and an interior (*bottom*) application.

Embossing or bumping the surface of copper also requires the use of less pressure than that used for steel. Its plastic behavior allows copper to be formed deeper on embossing rolls or with CNC-controlled equipment (Figure 5.24).

THE CAST FORM

Cast forms of copper alloys are common today in all industries, and in the realm of art, cast sculpture makes up a significant portion of the visual arts. Nearly every city in the United States has its share of cast bronzes made from the silicon–bronze alloys C87300 and C87600. In Europe and China silicon bronze is becoming more and more common as it replaces the alloy C83600, also known as Leaded Gunmetal or LG2. C83600 is still used extensively outside of the United States because of its ease of casting and finishing as well as the plethora of beautiful patinas that can be developed on the alloy.

Copper alloys can be cast using several common industrial techniques:

- Sand casting
- Permanent mold casting
- Die casting
- Investment casting
- Centrifugal casting

Copper alloys used in art and architecture are predominantly cast in sand. All of the copper alloys can be sand cast. The lost-wax method, also known as investment casting, is a sand cast method that has been in use for thousands of years. It is still commonly used in casting copper alloys for art and architecture. Other methods, including permanent mold casting and die casting, are used in industry for creating hardware and assemblies where quantities are significant. Centrifugal casting is a casting method also used to create parts for industry, but it is not as commonly used in the architectural markets. Centrifugal casting is good for creating repeated parts that have symmetry around an axis, such as cylindrical shapes. The spinning die receives the metal by means of centrifugal force. High accuracy and tolerance can be achieved.

Sand casting is the most flexible and economical of the casting methods. The use of rapid prototyping has influenced sand casting techniques by opening up the process to direct computer-generated design. Investment casting techniques as well are benefited from rapid-prototype generation of molds using low-melt wax or plastic printed using additive manufacturing techniques (Table 5.11).

There are four factors to consider both in the design and in the selection of alloy when casting copper alloys:

1. *Fluidity*. This is the ease with which a molten metal will flow through a cavity and into the mold. Good fluidity allows for fine detail and thinner sections and reduction of rejected castings due to incomplete mold filling. The foundry considers how far the metal must flow in the mold before solidification begins. Some alloys have better fluidity than others, and the foundry will help determine this.
2. *Shrinkage*. Shrinkage is a design consideration that should be considered in the design of the mold and the choice of alloy when working with any of the cast alloys. “Shrinkage” is the term given to a change in dimension as the liquid metal undergoes solidification in the mold. Different elements in the alloys solidify at varying rates and this will have a large effect on the soundness of the casting. As the casting begins to cool, it will solidify at a linear rate. Shrinkage and solidification are somewhat more predictable to the mold designer for castings that are linear in form as apposed to a wide bulky form. For complex shapes, predicting how the part will shrink is difficult. Always consider test runs to allow for the mold to be adjusted

TABLE 5.11 Comparisons of casting methods used on copper alloys.

Casting method	Relative cost	Process speed	Size	Surface quality	Dimensional accuracy	Common alloys
Sand	Low	Slow	Small to large	Poor to medium	Low	All copper alloys
Investment	Medium	Slow	Medium	Good	Low	Low-lead alloys
Permanent mold	Medium	Fast	Small to medium	Good	Good appearance	C85200 C85400 C85800 C86300 C90300 C99700
Die cast	High	Fast	Small to large	Good	Very accurate	All copper alloys
Centrifugal	High	Fast	Small to medium	Good	Very accurate	All copper alloys

to accommodate changes that may be encountered. Mold adjustments include adding risers and vents, tapering sections, and even adding chill bars to control where the solidification needs to occur first.

Alloy shrinkage is classified into three ranges—narrow, medium, and wide—each of which requires a particular mold design. A narrow shrinkage range solidifies from the walls of the mold inward and from areas of small mass to large mass. Alloys with a narrow shrinkage range require large risers, which keeps the thermal mass larger and in a position to feed into the casting. The medium range is the most forgiving of the three classifications and smaller risers are needed for medium-range alloys. It is difficult to achieve a good, sound casting for alloys with a wide shrinkage range. The edges of alloys with a wide shrinkage range will solidify at the same time as the rest of the casting, and the part will need to be designed so that the thermal masses are as uniform as practical.

3. *Dross formation.* "Dross" is the term given to the oxidized metal and other elements that float to the top and appear as a fine flaky film that solidifies on a casting. It can be removed by blasting the finish casting, but often pits or voids are left behind. In designing the mold for casting, keep the most visible, esthetic surface at the bottom of the mold. Dross is often due to adding the molten metal in such a way that turbulence and air is drawn into the mold as the metal is added.
4. *Temperature of pour.* A good design wants to avoid hot spots in the casting that will have an effect on the metallurgy and physical properties. Sharp, inside corners on the mold will cause the metal to stay hotter for longer.

Each casting method has its own unique characteristics and demands depending on design requirements, metallurgical characteristics, and production demands. Economy plays a large part in deciding which casting method to use. Rates of solidification of the metal is a major metallurgic consideration with each casting method. Sand casting has a slow freezing rate, while metal will freeze more rapidly when using permanent mold casting and die casting. This is due in part to the mold itself. Sand casting can provide insulation, keeping the molten metal warmer for longer.

Sand Casting

Sand casting has a low tooling cost, making it the most economical method of casting. Sand casting is straightforward: make a mold out of green sand and pour the molten metal into the cavity. Gravity pulls the molten metal into the mold and it flows to the edges.

Achieving a good surface finish, however, may be more difficult than in other, costlier methods due to the surface finish of the metal created by the sand mold.

Sand casting offers the ability to cast large sections as compared to other methods. This casting method is used in low-volume production processes and is particularly suited for custom, small-quantity features, such as the door handles shown in Figure 5.25.

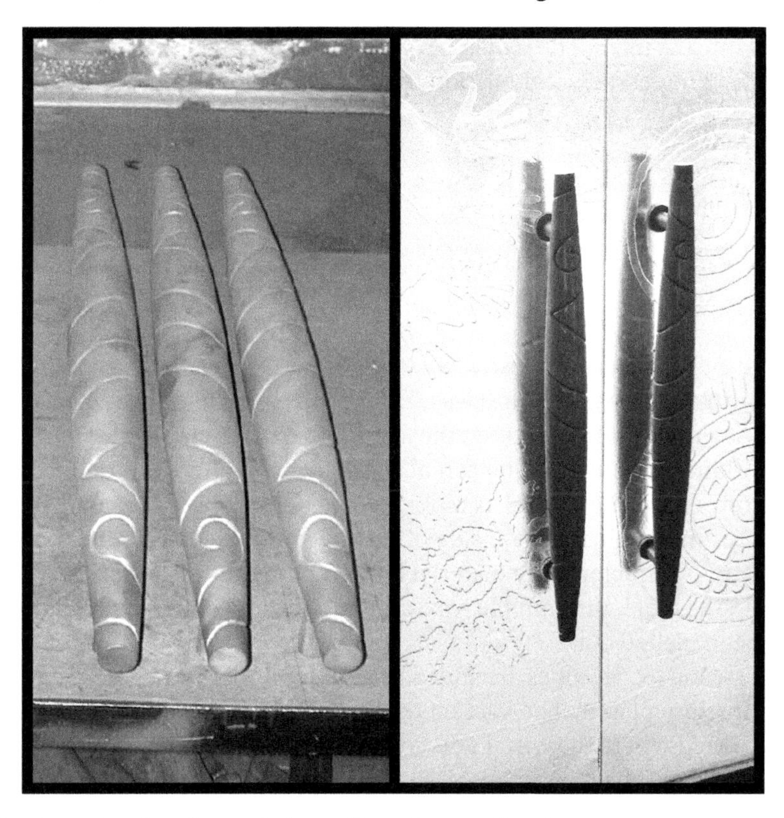

FIGURE 5.25 Sand cast door handles made from silicon–bronze alloy C87300.

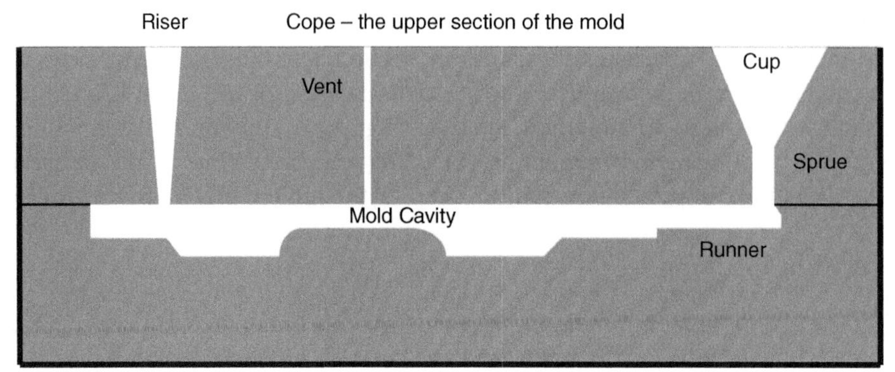

FIGURE 5.26 Cross-section of a typical sand cast mold.

Sand casting occurs in green sand. Cores are also of green sand. These are premade and baked to remove moisture, then set into the mold. In certain instances, the entire mold may be baked to produce a "dry mold" free of moisture, as moisture can lead to unwanted porosity. The molten metal is gravity-fed into the mold. The sequence of feeding the metal into the mold is designed by the foundry to establish a progressive solidification of the casting and reduce turbulence as the metal is added (Figure 5.26).

Investment Casting

Investment casting can produce very accurate designs with good dimensional control and good detail. The process is somewhat expensive due to the low run rates and the labor necessary to produce the forms.

Investment casting is also called "lost-wax casting" and has been in use for thousands of years. The French term *cire perdue*, meaning "lost wax," is another name for the method. In the investment casting method, first a model is made. Often the artist will make a clay model to represent the form that is to be cast. Additive manufacturing can also produce a highly detailed model from polymer material directly from a computer model. A mold is created over this model and allowed to dry, thus creating a negative of the original. The inside of this negative is coated with wax to the intended thickness of the casting. Refractory core material is often placed into the wax cavity to create a void. Nails are driven into the mold, through the wax, and into the core. These will hold the core in place once the wax has been melted out.

Wax vents, risers, and feeder tubes are added to the wax model while it is still in the mold; these are coated with refractory plaster. The wax model is now ready to be heated in order to melt the wax out and leave a void where the metal will go. The mold is then packed in refractory sand and prepared for the addition of the molten metal.

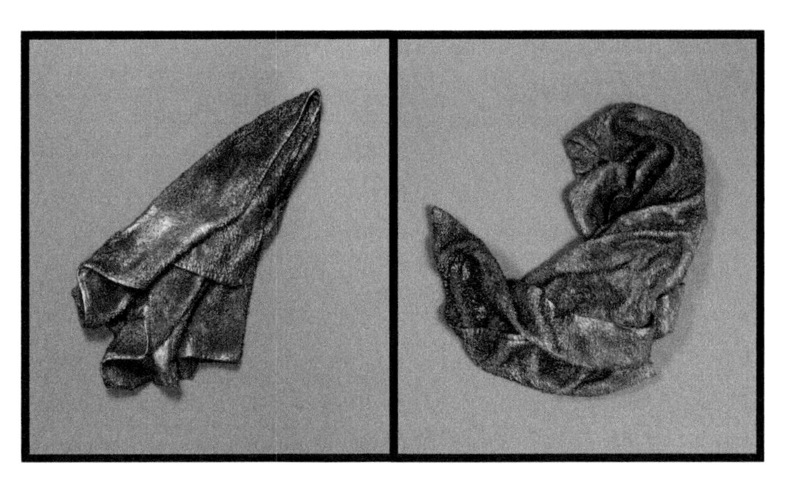

FIGURE 5.27 *Throwing in the Towel*
Source: by Michael Wickerson, Wickerson Studios.

Incredible detail can be achieved, such as that in the art shown in Figure 5.27. The fine texture of the towel is visible in these castings. The mold was pulled from an actual towel and carefully worked by the artist.

All alloys of copper can be cast using this method. Alloys containing large amounts of lead can have problems with the lead reacting with the calcium sulfate in the plasters used.

Permanent Mold Casting

The molds used in this casting method are made of metals such as cast iron or steel. They are used to cast the copper alloys over and over again. As in the sand cast process, this method uses gravity-fed molds. Dry sand or steel casting can make up the cores used in these molds, particularly when the finish is not critical. Permanent mold casting produces better surface finishes than sand molds, which tend to be grainy and lack smoothness. Dimensional tolerances are also tighter with permanent mold casting than with sand casting methods. Permanent mold casting falls between die casting and sand mold casting in mold cost and finishing production costs. Good detail is not easy to achieve with permanent mold casting. This technique is better suited to large quantities of simple cast forms needing a good surface.

Die Casting

Die casting is a process better suited to low-temperature metals. The metal is forced into the mold under pressure. The molds are often water-cooled metal dies.

TABLE 5.12 Factors to consider for various cast methods.

Factors	Sand casting	Investment casting	Permanent mold casting
Cost of process	Low	Moderate to high	Moderate
Production rate	Slow	Slow	Fast
Size of casting	Small to large	Limited size	Limited size
Shape	Complex	Complex	Less complex
Core	Sand	Plaster	Steel
Dimensional control	Good	Excellent	Good
Surface	Rough	Very good	Good
Porosity	Low	Medium	Low
Cooling rate	Slow	Slow	Slow to moderate
Grain	Coarse	Fine	Fine
Strength	Low	Low	Medium
Versatility	Excellent	Excellent	Very good
Repeatability	Poor	Poor	Good
Cleanup	Moderate	Minimal	Minimal

Die casting involves greater cost in the mold design and fabrication, but the surface finish achieved is superior. This casting method is best suited for multiple parts. Die casting requires placement of the molten metal into the die while under pressure. The surface finish of the copper alloy matches that of the steel die the metal is pressed into. Some post-trimming may be required, but cleanup is minimal.

Double die casting is another common method used to achieve a quality cast part. The part is first cast, then stamped to achieve tight dimensional control and surface quality. The process is well suited for automation, but since this method requires alloys to cool rapidly, alloys with tight freezing ranges should be used. Yellow brasses, manganese brass, and silicon brasses meet this requirement.

Casting is influenced by several factors that will determine the best method to use. Starting with the design of the part to be cast will help determine the most appropriate process, the alloy to use, and the expectations for the final finish (Table 5.12).

CHAPTER 6

Fabrication Processes and Techniques

. . . thus the rigid and inflexible will surely fail,
While the soft and flowing will prevail.

Tao Te Ching

INTRODUCTION

Ductility is the ability of a substance to deform plastically without fracturing or tearing. A defining characteristic of copper is its ductility. There are few metals that possess a similar ability to undergo shaping without tearing. Copper alloys, particularly the softer tempers, lack the anisotropic directionality that many other metals exhibit. For centuries, copper has been hammered and shaped to create useful forms by thinning and stretching the metal.

One measure of ductility is elongation of a material while under a load. Copper and many of its alloys, particularly copper–zinc alloys, can be stretched without breaking as loading is applied. Figure 6.1 shows a typical stress–strain graph of copper. The graph shows a flattened stress–stain curve; as load is applied, the copper article will elongate and undergo plastic deformation. The area under the curve represents the toughness of the copper.

Copper work hardens as it is shaped and molded. In repoussé and chasing metalwork, where the ductile copper is hammered and stretched, the copper will work harden with each blow. The hammering is repeated over and over again until very gradually the desired form is achieved. But there comes a point at which the continued hammering is resisted by the work-hardened metal and there is potential of tearing. Repoussé work-hardened copper is heated to a point below melting,

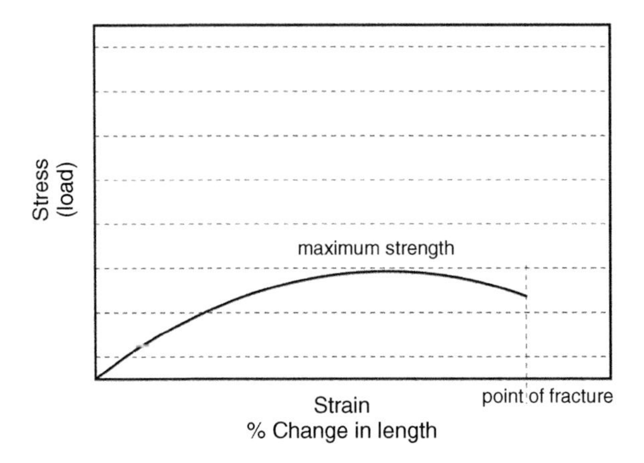

FIGURE 6.1 Stress–strain graph of copper.

then allowed to cool. This heating anneals the metal, and it becomes soft and ductility returns. Further hammering and shaping can now occur. This is repeated as each step takes the artist further in the shaping of the work.

As copper work hardens, the grains are stretched. This stretching realigns the crystal matrix as sections slip over one another and the copper elongates. Heating the copper causes the atoms to move about from areas of high stress to areas of lower stress. The crystal lattice relaxes, and further cold working can occur. The process can be repeated to a point.

The interesting part is that copper sheet does not display the antistrophic character that most sheet metal forms have. In wrought forms there are differences in physical characteristics along the length and across the length, but shaping with the grain and against the grain afford little differences in resistance. The hammered vase shown in Figure 6.2 is an example of the amazing ductility

FIGURE 6.2 Hammered copper vase.

FIGURE 6.3 Ancient Roman helmet hammered out of copper alloy.

of copper. Stretching it and conforming it to shapes by hammering, heating, quenching, and rehammering are part of an art form that extends back centuries.

Working with copper and copper alloys initially centered around castings to produce cast ornamentation and cast utensils. Helmets and shields were some of the first items using hammered copper (Figure 6.3). These offered some protection from clubs and spears made of wood and rock, but their real advantage was they could be straightened by rehammering, and if damaged, melted down and reshaped.

Some of the earliest swords were made of bronze, an alloy of copper and tin. This alloy was harder than copper alone and would take and hold an edge better. Most early swords were simply wood swords tipped with metal points. Bronze swords and daggers were cast and then forged and sharpened by early metalworkers.

The copper, bronze, and brass forms we use today all have roots going back to these early metalworkers and the designs they created. Here was a hard but pliable material that could be sharpened and strengthened by hammering and, once damaged, melted down and repurposed into a new item.

Copper gave early mankind the ability to work a material like nothing before it. By 2500 BCE many of the forming processes still used for metal today had been developed and were in use. Hammering, shaping, forming, and even welding were mastered by certain discrete regions where civilizations prospered and the metal was available.

FORMING

The ductility of the wrought forms of copper and copper alloys is a significant benefit to be exploited by the designer and fabricator. The condition of work hardening when the metal is cold worked is important to understand and appreciate. All wrought forms have a directionality, or a bias that dictates anisotropic differences in mechanical behavior. This directional bias is induced into the grains of the wrought metal as it is cold worked, stretched into long lengths of sheet, and recoiled. The same is true of bar and rod. Hot rolled plate also has a directional bias, but cold rolling stretches the grains and elongates the internal crystal structure.

In most applications directional bias is not a significant issue; the operator adjusts for the change in responsiveness in the forming operation. There is not significant springback in copper alloy forming, but there can be an increase in forming pressure and potential cracking when cold forming with the grain versus forming against the grain for some of the stiffer alloys, such as C28000. Figure 6.4 shows a variety of items made from copper, including a copper sink, copper shelving, and custom light sconces in which glass was blown into the assembled forms.

Forming parameters depend on alloying constituents, the cold rolled temper, and the direction or orientation of the bend. Additions of zinc, for example, will increase the strength and increase the rate of work hardening of the copper alloy. The higher the temper is, the more springback will be present and the more difficult the metal will be to form. All cold rolled sheets have an anisotropic, or directionality, imparted by the process of cold rolling. Cold roll anisotropy is induced in the long direction of the sheet or ribbon of metal.

FIGURE 6.4 Various fabricated copper items.

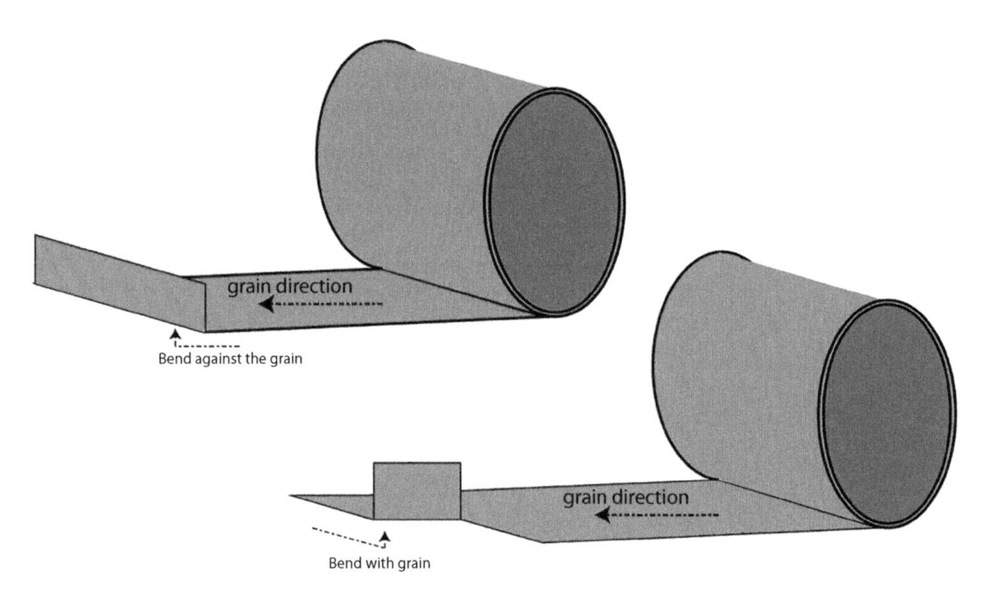

FIGURE 6.5 Bending with and against the grain.

Bending with the grain, as depicted in Figure 6.5, requires less force than bending against the grain. But with copper and copper alloys, the difference can be overcome with modern equipment and skilled operators, so it is not typically an issue. The dual phase alloys—those that have both alpha and beta phases in the crystal structure, which is a metallurgical condition that develops as alloying elements are added to the copper—are more prone to cracking when cold worked. Alloys such as C28000 (Muntz Metal) are dual phase copper–zinc alloys. Zinc is added to copper to around the 40% level. This alloy is difficult to cold work and can crack along tight bends. However, this alloy performs well when the metal is heated and the shape is formed through hot working.

V-CUTTING

V-cutting the back of copper alloy sheet or plate removes metal and leaves a small v-shaped cut partially through the material. This enables a tight fold to occur versus a rounded fold. Figure 6.6 shows the expected result of v-cutting.

Copper alloys can be v-cut using a method similar to that used for other metals; however, the back should be reinforced. V-cutting weakens the corner, and as the bend occurs the metal will undergo plastic deformation. You only have a single opportunity to make the bend. If it is folded, then flattened back out, subsequent refolding will likely crack the alloy at the point of the fold. The technique produces an intriguing appearance in thick sheet or plate forms of the metal, but it has structural limitations. The back can be reinforced with another angle of stainless steel or copper and bonded to the copper alloy by soldering or brazing a support piece or using very high bond (VHB) tape. It is good practice to test the connection and the reinforcement of the corner to ensure that no esthetic issues arise on the visible side. Figure 6.7 shows v-cut corners on 1.5-millimeter-thick C11000 panels at the de Young art museum.

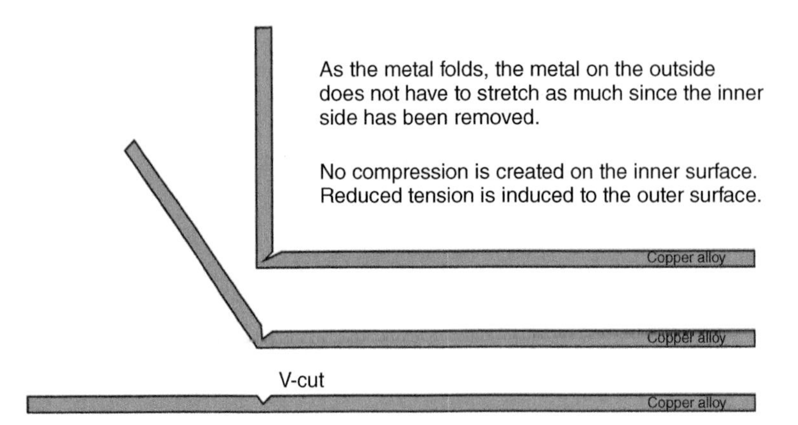

FIGURE 6.6 V-cutting copper alloys.

FIGURE 6.7 V-cut corner on panels at the de Young museum of art.

FIGURE 6.8 Shaping copper alloys.

Copper was made for the fabricator. The ability to shape the surface of wrought sheet is unmatched. It's as if you combined clay with steel. Pressing intricate shapes into the surface is what the metal was created for. As depicted in Figure 6.8, copper alloys can be hammered into forms and exploded into curved bubbles or rolled into repeatable waves.

Copper alloys will work harden as cold work is performed on the wrought forms. The work hardening makes subsequent shaping more difficult. The amount of work hardening that can be experienced depends on the alloy. The higher the zinc content, the more rapid the work hardening from cold working the metal. While the C11000 alloy will work harden at a relatively slow rate, the nickel–silver alloy C75400 and the phosphor–bronze alloy C52100 will work harden rapidly and can more than double in tensile strength from fully annealed levels.

Zinc, when alloyed with copper, induces more strength into the copper: as zinc is added, the strength of the alloy is better than that of copper alone. The alloys of copper and zinc that are in solid solution up to around 30% zinc have improved strength but retain ductility. Additionally, these alloys have greater elongation when under load than copper, and work harden at a lower rate than copper. Moreover, when stressed, necking occurs at higher strain rates.

As the zinc increases to 40%, a phase change occurs and both alpha and beta crystal phases develop. Alloys with this level of zinc are excellent hot working alloys and have superior machineability, but they are very poor cold working alloys and can crack under simple forming operations. They can be curved in plate rolling operations but will show significant resistance if the radius is tight. In thick wrought forms, shaping by brake forming will require skill to avoid cracking at 90° bends.

The graph shown in Figure 6.9 indicates the changes in yield strength of a H01 (quarter-hard) temper wrought sheet, as the zinc is added to alloys C21000 to C28000. At around the 40% zinc that is in alloy C28000, it reaches a maximum. Naval Brass (alloy C46400) has small amounts of lead

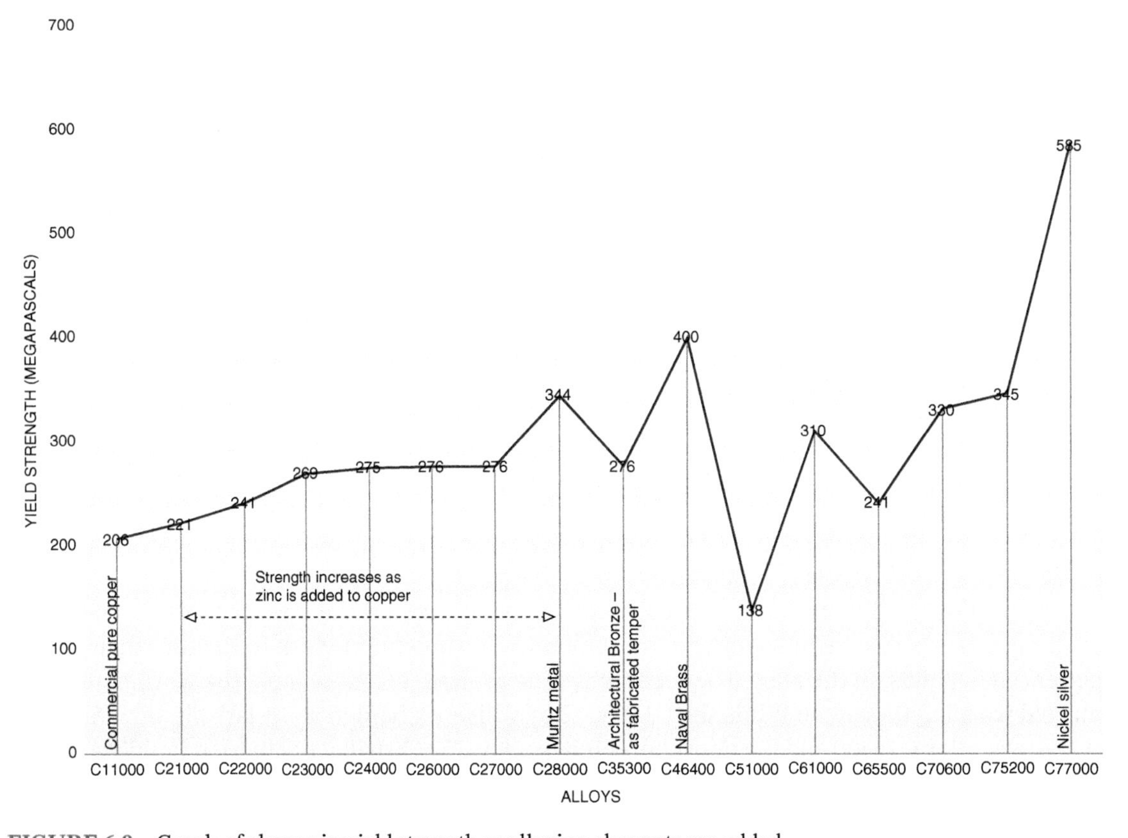

FIGURE 6.9 Graph of change in yield strength as alloying elements are added.

and tin added to basically the same alloy constituents as C28000, yet the yield strength increases significantly. The temper shown in this graph for alloy C46400 is H02, resulting in higher yield strength than H01.

Yield strength increases significantly as nickel is added. The copper–nickel alloys C70600 and C75200 have a yield strength much higher than commercially pure copper and the nickel–silver alloy C77000 possesses the yield strength of steels. Alloy 35300, known as Architectural Bronze, is an extruded form and therefore the temper is an as-manufactured temper. Alloy C51000, phosphor bronze, shows a low yield strength when provided in H01 (quarter-hard) temper. It jumps up considerably as further cold working raises the temper to H02. At this point the yield strength of C51000 is similar to C28000 at an H01 temper.

The alloy and the temper play a major role in the strength characteristics of copper alloys.

True bronzes, alloys of copper and tin, are rarely used in art and architecture. The cast alloy C87400, commonly known as silicon bronze, is used in numerous cast sculptures and contains no tin whatsoever. The wrought form of this alloy, C65500, has good strength and ductility. Its yield strength is similar to that of alloy C22000, Commercial Bronze.

There are occasions when the phosphor–bronze alloys C51000 and C54400 are used in wrought forms of plate. These alloys contain small amounts of tin, up to 5%. The alloy can be cold worked but its strength increases rapidly, making subsequent forming more difficult. Annealing the alloy will lower the temper and allow for further forming.

Copper alloys that contain aluminum (C61000, C61300, and C61400) are known as aluminum bronze. These have good strength and hardness. These alloys will form better than the high-zinc alloys.

Copper–nickel alloys contain an appreciable amount of nickel. Copper and nickel are two metals that are miscible with each other in any quantity. Copper–nickel alloys are ductile and very resistant to corrosion, particularly in seawater. They were once used extensively on the external sides of ships below water to act as protective biocides against barnacles and worms.

The nickel–silver alloys C75200 and C77000 contain zinc as well as nickel. These are available in limited sizes, but they can be cold worked. They are very strong and hard alloys. As they are cold worked, the temper increases, as does the tensile strength.

The ductility of copper and copper alloys allow for tremendous flexibility of shape and form. Used as a thin cladding for surfaces, they will conform to the underlaying shape as if they were painted on the surface. Figure 6.10 shows three examples of thin copper material conforming to unique shapes and surfaces.

FIGURE 6.10 Copper and copper alloy–clad surfaces.

CUTTING COPPER ALLOYS

Copper and copper alloys can be cut by all conventional cutting equipment. The yield strength of copper and most copper alloys is lower than that of steels. Most of the copper alloys are also softer than many other metals. This makes cutting with steel shear blades and saws more effective. The modern cutting methods of waterjet and laser are common means of cutting copper alloys. Standard plasma will cut copper but will require edge cleanup. The high-definition plasma cutters will rapidly cut through copper alloy material and produce an edge that requires less cleanup. Knowing how the different cutting methods will respond to copper alloys is important. Not all lasers will effectively cut copper, and some can be damaged by the reflectivity of the molten metal. Plasma can melt copper to make the cut, but the rapid cooling can redeposit the molten metal as a coarse oxide along the edge. Table 6.1 compares the cutting methods used on copper alloys.

Shearing and Blanking

Shearing, slitting, and even hand shears will cut thin copper and copper alloys with little effort. These tools are mainstays in the sheet-copper fabrication shop. Shearing copper alloys produces a straight-line cut as the metal is sliced in a way similar to the way scissors cuts through paper.

TABLE 6.1 Cutting methods used on copper alloys.

Cutting method	Form	Thickness	Speed	Subsequent operations
Shearing	Sheet, plate, wire	<5 mm	Fast	Accurate straight-line cut. Foil sizes are cut with a shear, similarly to paper.
Saw cutting	Plate, bar, tube, extrusion	All forms and thicknesses except foil	Slow	This operation is slow and less accurate than other methods.
Waterjet	Sheet, plate, bar, extrusion, tube	>0.25 mm	Slow	Thin sizes may shape from pressure. Accuracy is superior.
Laser	Sheet, plate	<4 mm	Fast	Very accurate. Can shape from heat.
Plasma	Sheet, plate	<3 mm	Medium	Accurate but can shape from heat. Cleanup required to remove the burr produced along edge.
Machining	Plate, extrusion, bar, tube	>1.5 mm	Slow	Very accurate. Will need some post-finish work to remove machining marks.

Shearing is an operation involving two sharpened and hardened steel blades. One blade is fixed and the other is brought down with such force that a metal strip undergoes severe plastic deformation to the point that it fractures at the surface line where it contacts the shear blades. This fracture propagates through the metal. When there is proper clearance between the cutting blades, the crack that propagates penetrates only a portion of the thickness. One crack will form at the top blade and another crack will form at the bottom blade. These cracks meet near the middle to provide a clean fracture line.

Blanking involves a similar process. A tool with a sharpened edge comes down on a die, and the interaction slices through the copper alloy creating a blank form in the middle of a sheet or strip of metal. The balance is gathered and recycled. The blank is transferred and used in subsequent forming operations. Blanking is used when multiple forms of the same dimension are stamped out, whereas shearing is normally used for one-off operations or low-quantity operations.

Saw Cutting

Wrought and cast copper alloys can be cut with cold saw blades. Thick copper alloy material is often saw cut. Heat buildup is minimized because the copper will dissipate the heat generated during the cutting process. A cold saw is a steel circular or band saw. The tip of the blade can be coated in tungsten carbide, synthetic diamond, or titanium nitride. Saw cutting can induce heat into the copper alloy, causing localized distortions along the edge if the saw is not cutting properly or if it is worn out. It is good practice to conceal a saw cut edge in a cover plate or folded under in a lap joint. It is very difficult to achieve an acceptable saw cut edge in thin copper alloy sheet. Extrusions, bar, and tubing are often cut with a saw. The edge usually requires some re-dressing to remove burrs and soften the cut.

When saw cutting copper alloys, use saws with coarse teeth on thick and hard alloys and fine teeth on thinner forms of copper alloys.

Waterjet Cutting

Waterjet cutting of copper alloys is an excellent method of creating intricate shapes and forms from sheet and plate material. Thin copper alloys can be waterjet cut by leaving tabs to suspend the parts and keep them from bending from the force of the jet of water and abrasive in the water. Cutting with water can achieve fine detail, such as the detail of the leaf art form and grille cover shown in Figure 6.11.

On thick copper alloy plates, waterjet cutting is the preferred means to achieve intricate forms. Waterjet cutting is a means of accurately cutting thick metal plates to produce inserts—such as the brass inserts as shown in Figure 6.12—for inserting in terrazzo flooring or wooden walls to create decorative features.

Waterjet cutting does not induce any heat into the copper parts being cut nor does it interfere with the application of patina or statuary finishing. The gate shown in Figure 6.13 was manufactured from 3-millimeter-thick C22000 alloy material and assembled with a 3-millimeter stainless steel waterjet plate. The C22000 alloy was given a dark statuary finish to contrast with the stainless steel insert plate that adds to the overall strength.

Powerful waterjets can cut all metals accurately, but there are several things to keep in mind when cutting copper and copper alloys. For one thing, careful handling of the finished material to and from the bed is critical. Most waterjets use steel slats to support the material being cut. These slats are often jagged from previous cutting operations. Transferring finished or semifinished plates to or from the bed can scratch and mar the surface of the copper or copper alloy.

FIGURE 6.11 *Left*: Decorative leaf form by Reilly Hoffman. *Right*: Decorative grille cover made from the C22000 alloy.

FIGURE 6.12 Waterjet cut thick brass inserts.

FIGURE 6.13 Gate with waterjet cut angels.

Drilling of these metals should also be conducted carefully. Waterjets initially drill a hole through the metal sheet or plate. This initial force can be significant and shape or bend small parts. Because the metal surface is soft, the initial drill force can etch the surface around the initial point of entry creating a frosted halo.

Laser Cutting

With the advent of the powerful fiber laser, laser cutting has become an invaluable means of cutting copper and copper alloys. Copper has a different light absorption rate from other metals and thus cutting it requires a different wavelength and higher power. Earlier CO_2 lasers did not have enough power and molten copper would reflect the light back into the lens, causing significant damage to the laser. A fiber laser doesn't deliver a laser beam in the same way that a CO_2 laser does. Instead, power is delivered though a glass fiber. The glass fiber is doped with rare-earth elements and can support high kilowatt levels.

As compared to the waterjet method, one drawback of using a laser is that it delivers heat energy to the cut; for some intricately cut patterns, this can warp and shape the copper alloy. On a piece with very intricate and extensive cutting, warping can interfere with the cutting head and stop the machine. A warped sheet may require post-flattening of the cut form or part. Figure 6.14 shows

FIGURE 6.14 Laser cutting in *Fuga*
Source: by artist Jan Hendrix.

some extremely intricate cuts on 3-millimeter-thick alloy C22000. The artist, Jan Hendrix, designed these incredible sculpture forms to be cut from copper alloy for strength and then chrome plated to achieve a mirror-like appearance.

The edge produced from laser cutting is clean and precise. There is a slight halo of color that will occur at the cut edge from localized heating during the cutting process. Figure 6.15 shows a laser cut art form designed by artist Jason Pollen. The art was cut from alloy C11000 and finished with Dirty Penny. The copper is 2 mm thick. When the holes or shapes of cuts are small and tight some re-fusing of the hot copper alloy can occur at the cut, locking the cut out into the cut shape and requiring secondary means of removal.

Fiber lasers will cut copper alloy sheet of 3 mm and less with little difficulty. As with waterjets, the beds are often steel slat, or possibly copper slat. In any event, the molten material created by the laser cutting can redeposit onto the slats, making them jagged and sharp. These will scratch the copper if care is not taken in the removal of the cut parts from these jagged points.

FIGURE 6.15 Laser cut artwork
Source: by artist Jason Pollen.

Plasma Cutting

Copper alloys can be cut with plasma cutting equipment. Plasma cutting is typically performed on two-dimensional CNC-controlled tables. The copper alloy plate or sheet is set onto a steel lattice and a high-energy plasma beam cuts through the metal. Plasma cutting involves creating electrically charged ionized gas and forcing the gas through a small orifice. An electrical current is generated from a remote power source that creates an arc from the copper work piece, which is given a positive charge, and the plasma gas, which is given a negative charge. Plasma cutting can also be performed by hand using portable plasma stations. Figure 6.16 shows hands cut out of 1.5-millimeter-thick copper sheet and the final artwork. The edges required cleaning up to remove the coarse dross left along the cut line where the melted copper attached back to the edge.

The gas used for cutting copper is nitrogen or an argon–hydrogen mix. Carbon dioxide and oxygen gases can be used to create the plasma jet, but these will create a darkened cut as the copper alloy edge oxidizes.

Temperatures in the high-velocity plasma jet can be as high as 22,200 °C, as this highly charged gas melts the metal and blows it away. There is not a lot of heat transferred to the copper, but if there are a number of piercings you can expect warping. Piercing with plasma is similar to piercing with a laser. The energy must "drill" a hole through the metal, and as this concentrated heat sits on the surface for just a moment, the copper will absorb it.

High-definition plasma cutting systems reduce the heat-affected zone and produce a fine-cut line in thin copper. Handheld and less sophisticated systems will have a larger kerf and more oxidation on the edge. The kerf, which is the term used to describe the edge of the cut or the cut itself, will be rougher than that produced with a waterjet or laser. There can be a redeposit of molten metal along the kerf. This redeposit is rough oxide that will need to be removed by filing or sanding.

FIGURE 6.16 The plasma cut copper artwork *Hands of Man* by the author.
Source: Photo courtesy of the author.

MACHINING

Pure copper does not machine particularly well, as the rapid cutting tools used on machining centers produce long curled tubes that can gum up the tooling. Chatter can form as the more ductile pure copper flexes under cutting loads. Adding small amounts of tellurium, sulfur, or lead will cause the copper trailing to break off into small shards, making the machining of these alloy forms more effective.

Most of the copper alloys considered for use in art and architecture are similar to copper when it comes to machining. The harder, two-phase alloys that possess both the alpha and beta crystal structures are better adapted to machining processes. Copper–aluminum alloys machine similarly to the way that steels machine. The hardness of these alloys will cause rapid tool wear. The copper–nickel alloys are very difficult to machine because of rapid tooling wear induced by the nickel.

Machining is a broad term, but for the most part it involves the selective removal of small amounts of material by passing a special, rapidly rotating tool over the surface. Table 6.2 discusses the machineability of various alloys that are used in art and architecture. Alloys with good machineability are often referred to as "free-cutting" alloys. The industry assigns a number to metals reflecting their machineability, with 100 being highly machineable.

Lead added to alloys assists in the lubrication of the cutting action of the tool and causes the trailing to be small and to break away. Leaded nickel–silver alloys will machine better, but the tooling will wear more rapidly than it will for steels because of the nickel content. Many industries machine fixtures, pumps, and special hardware that require machining to tight tolerances. Small parts and fittings are commonly machined from various copper alloys (Figure 6.17).

There are copper alloys designed specifically for the purpose of machining. One, commonly known as "free-cutting brass," is a copper alloy consisting of 39% zinc and 3% lead. This alloy, C35300, requires lower cutting forces than aluminum and can be machined more effectively than alloys of steel or aluminum.

As the machining tool addresses the block of C35300 alloy, small shards are quickly and efficiently removed. This reduces post-finishing and allows for intricate designs to be produced on end or table mills. Machined copper alloys are used in the fabrication of hardware, marine fittings, and valves, and in many other areas where excellent corrosion resistance and durability is desired.

The Museum of the Bible art panels designed by Larry Kirkland and shown in Figure 6.18 were machined from copper alloy C35300. Initially casting was considered as a means of producing the art, but casting would have required numerous unique physical molds, the cost would have been higher, and significantly more time would have been required for fabrication. Machining, on the other hand, is a production process that is accurate and rapid, and was a way to "carve the artwork" out of a metal plate—in a way, a kind of reverse rapid prototyping. Instead of using physical molds, the design was created in three-dimensional computer code and metal was cut away from a solid 25-millimeter plate of brass to reveal the lettering. All the material removed was recycled. Figure 6.18 lists the steps used to machine the 25-millimeter-thick block of alloy C35300.

A three-dimensional computer model of each letter was created, and instructions were provided via CNC to an advanced milling machine, where the lettering was cut out of 25-millimeter-thick

TABLE 6.2 Machineability of copper alloys.

Alloy	Machineability	Point of discussion
C11000	20	Copper has poor machineability; it gums up tooling.
C21000	20	High-copper alloys possess poor machineability.
C22000	20	High-copper alloys possess poor machineability.
C23000	30	The added zinc makes this alloy harder, but still poor machineability.
C24000	30	The added zinc makes this alloy harder, but still poor machineability.
C26000	30	The added zinc makes this alloy harder, but still poor machineability.
C28000	40	This alloy is harder, but still poor machineability. C35300 is similar but machineable.
C35300	**90**	**A leaded alloy that is considered free-machining. Excellent machineability.**
C37000	70	Good machineability due to small amounts of lead.
C46400	30	A hard alloy, but still poor machineability.
C51000	20	Poor machineability.
C61000	30	Poor machineability.
C65500	30	Poor machineability. Nickel content wears out tooling.
C70600	20	Poor machineability. Nickel content wears out tooling.
C75200	20	Poor machineability. Nickel content wears out tooling.
C77000	30	Poor machineability. Nickel content wears out tooling.
C79800	50	Improved machineability due to the added lead.
C83600	**90**	**A cast alloy with excellent machineability. Considered free-cutting.**
C84400	**90**	**A cast alloy with excellent machineability. Considered free-cutting.**
C85200	**80**	**A cast alloy with excellent machineability. Considered free-cutting.**
C86300	60	A cast alloy with good machineability.
C90300	50	A cast alloy with moderate machineability.
C97300	**70**	**A cast alloy with excellent machineability. Nickel wears out tooling.**

Bolded numbers are considered free-cutting alloys.

FIGURE 6.17 Machined copper alloy parts.

FIGURE 6.18 Machining, texturing, and finishing of panels at the Museum of the Bible.

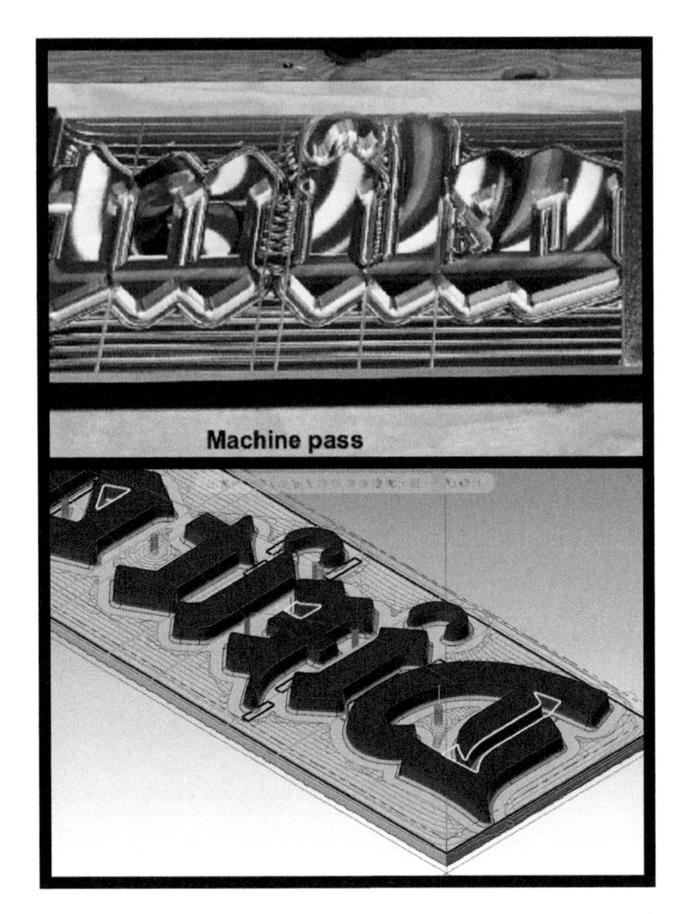

FIGURE 6.19 *Top*: Initial mill marks on the metal surface. *Bottom*: Computer image of the machining passes.

plates. The top image in Figure 6.19 shows the initial mill marks on the surface, while the bottom image is a screenshot of the movement and tool paths of the machining tool.

For the entryway of the Museum of the Bible, the artist Larry Kirkland chose a copper alloy that would, after texturing and oxidizing, emulate the way the lettering dies used in a Gutenberg press of old would look. The first two pages of Genesis, written in old Latin script, were chosen for the C35300 panels flanking the door. As with a real press, they had to be mirror images. Alloy C35300 was chosen.

The machined plates were needle hammered around the letting to create a special texture, then darkened using a copper–sulfide statuary solution and highlighted to mottle the surface. As Michelangelo once said, "The marble not yet carved can hold the form of every thought the greatest artist has" (Figure 6.20).

FIGURE 6.20 Machined lettering at the Museum of the Bible, designed by Larry Kirkland.

SOLDERING, BRAZING, AND WELDING

We think of soldering, brazing, and welding processes as inventions of modern age. It cannot be argued that modern welding technology is one of the most advanced and innovative areas of metalworking. But the end result—two discrete metal forms joined together to form one—is a form has been practiced for more than 5000 years. In 3000 BCE the Sumerians, one of the most prolific early Bronze Age cultures, made swords of bronze that had sections joined by hard soldering. The Sumerians even used a flux to braze parts together. It is interesting to think that the processes of joining metal together have been built upon for over 5000 years, as advancements in technology drive the technique to new levels.

Soldering and Brazing

Soldering and brazing are common forms of joining copper and copper alloys. The processes of soldering and brazing are similar. Brazing is sometimes referred to as "hard soldering," while soldering is occasionally called "soft soldering." The difference lies in the heat and filler metal used. The general rule is:

Soldering	Low melting point fillers	Less than 450 °C
Brazing	High melting point fillers	Greater than 450 °C

In each method, only the filler metal and not the base metal is melted. Soldered and brazed joints are strong and can transfer stress across them, but they are not as strong as the parent material.

Soldering and brazing are forms used to join thin copper alloys together or to seal the edges and seams of assembled units with metal filler material. Soldering and brazing form a metallic joint between two metals using thermal processes that melt the filler metal at temperatures below the base metal's liquidus temperature. The metals being joined by the soldering process can be the same or different metals. The filler metal makes a metallurgical bond with the base metals, joining them together. Soldered and brazed joints have the ability to conduct electrical and thermal energy.

Soldering uses filler metals that have low melting points, such as lead, tin, and antimony. Brazing uses filler alloys with higher melting points, such as silver–copper–phosphorus alloys.

Soft solders generally in use today include both leaded and lead-free varieties. Leaded solders are tin–lead solders with very low melting points. The lead solder combination of 63% tin and 37% lead is considered a eutectic alloy (that is, they both dissolve completely in one another). Because of this, there is a specific temperature at which they transform from solid to liquid. Other lead solders liquify at a range of temperatures (Table 6.3).

Leaded solders are being phased out because of the hazards associated with lead exposure. These solders were in common use for decades and still can be found in common use in the United States. They have been discontinued for the most part in Europe and Japan. The difficulty in replacing them is in the ease of use. Lead has an excellent wetting ability; that is, it breaks the surface tension in the mix and allows the molten metal to flow.

Lead-free solders have a higher melting point than leaded solders (Table 6.4). These solders are alloys of tin, silver, antimony, copper, and sometimes indium and bismuth.

There are several versions of lead-free solders in common use today. They are somewhat more difficult to apply. Good clean joints and a compatible flux are needed to make a sound solder joint with these solders.

TABLE 6.3 Leaded solders.

Solder		Melting point	
63/37	63% tin, 37% lead	183 °C	360 °F
60/40	60% tin, 40% lead	183–190 °C	360–375 °F
50/50	50% tin, 50% lead	183–215 °C	360–420 °F
2/98	2% silver, 98% lead	304 °C	579 °F
5/95	5% silver, 95% lead	304–370 °C	579–698 °F

TABLE 6.4 Lead-free solders.

Solder		Melting point	
95/5	95% tin, 5% antimony	240 °C	464 °F
SAC	95% tin, 4% silver, 0.5% copper	217–220 °C	422–428 °F
97/3	97% tin, 3% copper	227–300 °C	440–572 °F

Producing good solder joints with soft solders relies on good, clean surfaces being joined. The flux used is very critical as well, particularly for the lead-free solders. The time between flux application and solder is critical. Some of the lead-free solders are available with flux cores to facilitate application at time of solder.

The design of the joint is also critical. A soft solder has relatively low mechanical strength, so it is critical not to overstress the solder joint. Shear strength is low, so it is important to design the joint so as to not place undo shear stress across a soft solder joint.

Additionally, a good continuous heat source is needed to keep the metal joint being soldered at a consistent temperature. This will allow the solder to flow. It is very critical to clean the joint after soldering with fresh water to remove and neutralize any flux. Figure 6.21 shows soldered joints in a copper roof designed to act as a permanent metal seal to keep moisture out.

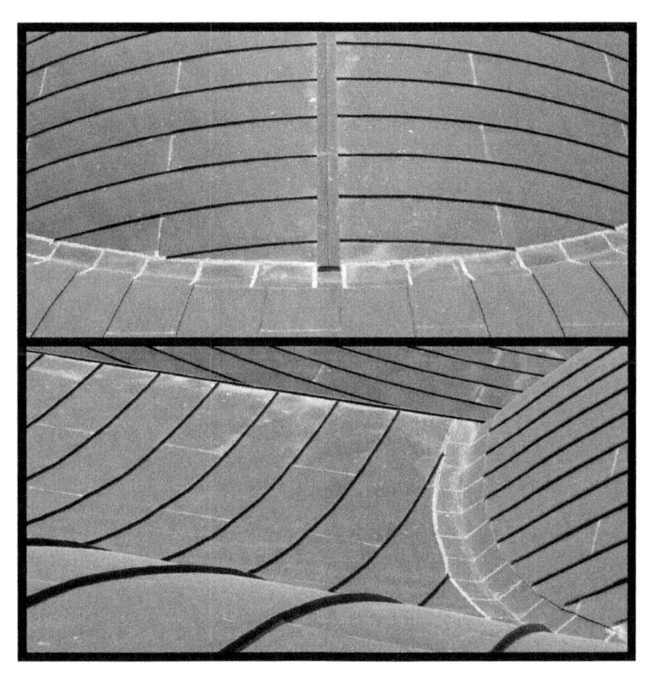

FIGURE 6.21 Soldered joints in a copper roof.

TABLE 6.5 Common brazing alloys.

Brazing filler		Melting point	
72/28	72% silver, 28% copper	790–835 °C	1435–1535 °F
80/15/5	80% copper, 15% silver, 5% phosphorus	829–860 °C	1525–1575 °F
91/2/7	91% copper, 2% silver, 7% phosphorus	643–788 °C	1190–1450 °F

There are several copper alloy brazing filler metals. Some are designed for brazing in a vacuum or with an inert gas. Typical brazing filler metals contain silver along with the copper, with phosphorus included to aid in fluxing the base metal surface.

Copper and silver are eutectic at mixtures of 28% copper and 72% silver. Alloys of this mix can be used as a brazing material, but the more common materials use a copper–silver–phosphorus mix often referred to as "sil-fos." Table 6.5 shows a few of the common alloys used in brazing.

The major benefit of brazing is the strength achieved at the joint. These joints have much higher mechanical strength due to the filler material and the metallurgical bond developed. Brazing also exhibits better corrosion resistance than soft-soldered joints because the metals are more compatible. The difficultly is they require significantly higher temperatures, and therefore thermal movement needs to be taken into consideration. Color transformation from the heat also needs to be dealt with. Figure 6.22 shows two examples of artwork in which brazing were used. One uses it on the joints of a blown-glass cage form, the other on the joints of a form of copper wire and strip.

Brazing is used on thicker copper alloy material, while soldering is generally used on thinner material. All copper alloys can be soldered or brazed, but soldering is generally considered for copper roofing, guttering, and similar thin fabrications where strength is not critical but sealing the joint from moisture infiltration is.

FIGURE 6.22 *Left*: Brazed copper with blown glass. *Right*: Brazed copper wire and strip.

Brazing produces a strong joint, one that is not as strong as a welded connection but one that seals as well as transfers stress across the metal.

Keys to a good solder or brazed connection:

- Thoroughly clean the joint
- Use a compatible flux
- Use an efficient joint design, keeping the gap between surfaces tight
- Use a proper heat source
- Clean up, neutralize, and flush the joint

From a corrosion standpoint, the last item—clean up, neutralize, and flush the joint with fresh water—is critical. The fluxes used are often acids and these need to be brought to a neutral pH and flushed from the surface. Fluxes are necessary to prepare the base metal surface to receive the filler metal. Flux residue can generate corrosion cells if left on the surface. A clear sign of acid remnants is show of green patina corrosion products developing around the joint after a short period of exposure to the atmosphere. Fluxes are designed to activate below the temperature of the melting point of the filler metal. They must remove the existing oxide on the surface of the copper alloy in order to allow the filler to thoroughly wet the joint.

Fluxes commonly used for soldering copper alloys contain zinc, ammonium chloride, or phosphoric acid. Flux residues will corrode the copper alloy if they are not adequately removed. They come in paste or liquid forms.

Noncorrosive fluxes that use a zinc bromide are available, and although they are not as effective as corrosive fluxes, in certain instances they are preferred. In particular those instances where flushing the flux residue from the joint is difficult.

Brazing fluxes operate at higher temperatures. They contain boron compounds such as boric acid and sodium tetraborate (borax compound). These are also available in paste form. It is important to remove all excess flux after brazing to prevent corrosion of the metal surface.

The ideal soldered joint or brazed joint achieves a full seal. To achieve this the filler metal must be drawn into the joint and wet the surfaces sufficiently to achieve a bond. Figure 6.23 shows two types of joints used on thin copper alloy material. The simple lap seam shown at the top is weak,

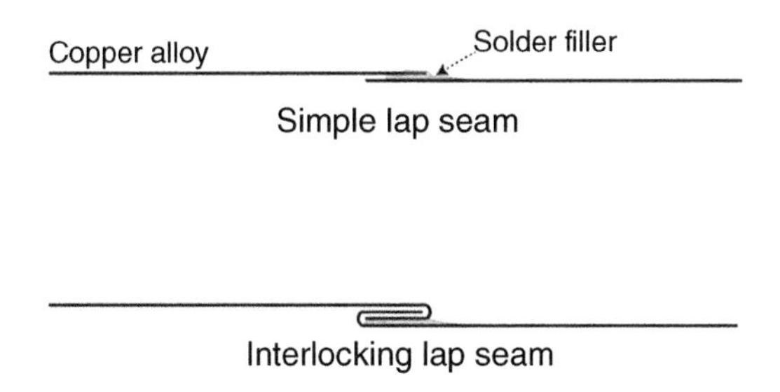

FIGURE 6.23 Two solder joints on thin sheet material.

with the solder expected to bridge the metal to provide needed stiffness and strength. The interlocking lap seam shown at the bottom joint provides strength to the joint, and when sealed with the filler material creates a sound joint that can accommodate significantly more stress than the simple lap seam.

For a soldered or brazed joint to be effective, the filler material must bond with the surfaces of the copper alloy. This is referred to as "diffusion." The clean metal surface is heated to above the melting point of the filler material. As the filler material touches the hot base metal it liquifies and is pulled into the joint by means of capillary pressure, filling the space as it mechanically keys into the surfaces. The metal is allowed to cool and the filler solidifies, making a metallurgic bond between the filler and the surface of the base metal. The joint must be tight to facilitate the forces of capillary action.

When applying heat to the base metal to be joined, it is important to heat the entire joint. The heat first melts the flux, which wets the area to be joined. This activates the surface of the base metal. Next the filler metal is melted and drawn into the joint, effectively filling the gap between the base metal pieces.

Welding

Copper and copper alloys can be welded by all conventional means used on other metals. There are differences, however, that must be considered for effective welding to occur. Copper and most of the alloys have high electrical and thermal conductivity. For welding this means the heat of fusion is conducted away more rapidly and incomplete welding of the base material may occur, leading to higher porosity around the weld.

In copper alloys, in particular those that contain high levels of zinc, the zinc will vaporize during the heat of welding and fume from the molten metal. This can deplete the zinc in the weld area, leading to color differences when finishing processes are undertaken. The fumes are also hazardous and proper venting and protection should be practiced.

Copper alloys get their strength from cold working. The intense heat generated from welding will produce a weaker assembly. Thickness and shape may be needed in the design to overcome this decline in strength. The high thermal expansion coefficient of copper will also increase weld distortion concerns. These too, require proper design and assembly to overcome.

For all copper alloys it is important to begin with a clean, oxide-free surface before attempting to weld. Copper oxides will hamper welding in a way similar to soldering.

Beginning with an oxide-free surface is the first step in successfully welding materials to be joined.

As welding proceeds, the use of special fluxes or gas atmospheres around the weld must be utilized to prevent the oxide from returning during the welding process. This would lead to porosity in the weld.

Preheating the parts to be joined will reduce the issue of thermal conductivity and allow for lowering the heat input into the weld joint and in interpasses. This will help reduce cracking as the metal cools.

Welding Copper

Copper can be welded using arc welding techniques. Those techniques using gas shielding, such as an argon gas, provide the needed shielding to keep oxides from forming. When welding copper, consider using the gas metal arc welding (GMAW) or the gas tungsten arc welding (GTAW) process. Other methods can be used, but achieving good results is more difficult. Note that with copper, the finish will be consistent and oxidation—either naturally formed or artificially formed—will cover the metal well. The images at the left in Figure 6.24 show examples of welding C22000 seams and edges, while image on the right shows copper bar being welded for a handrail.

Thin copper can be welded. The difficulty comes with achieving a smooth weld without weld distortion. Figure 6.25 shows a very difficult weld being made on 1.2-millimeter-thick copper. The dimples help to add stiffness to the elbow, allowing a very fine weld to be laid down to join the copper. Copper rod was used as the consumable and the resulting joint was post-polished.

Welding Brass Alloys

There are several issues to overcome when welding brass alloys. Esthetic challenges can develop when matching the color of the alloy at the weld. The constituents in the copper alloy, specifically zinc, will vaporize and there is a need for using fillers to help put some of the zinc back. The better solution is to attempt to design the weld in a less visible location. Statuary patinas applied to the weld area will also react differently and appear different. Patinas are not paints and will not cover.

FIGURE 6.24 *Left*: Welding of C22000 plates. *Right*: Welding of a handrail using the C11000 alloy.

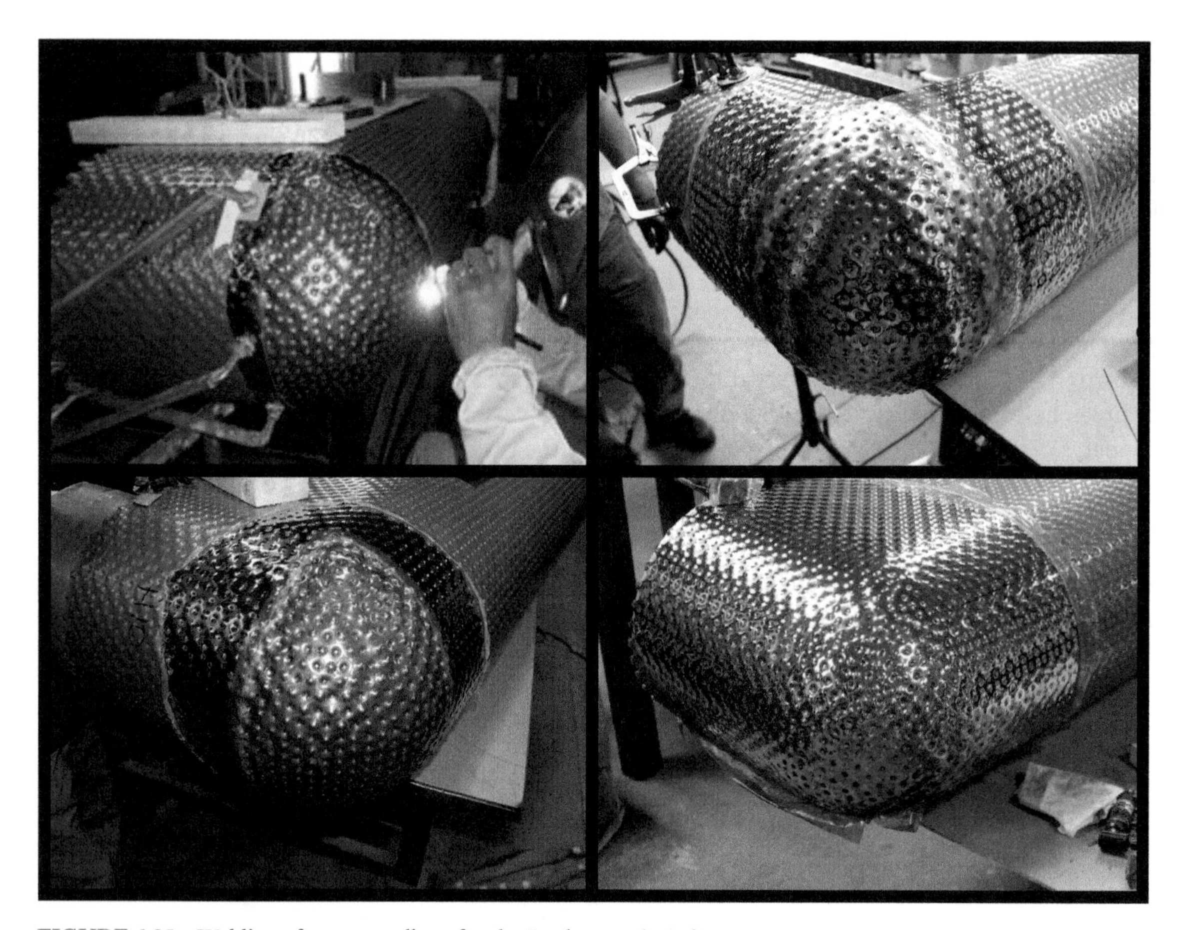

FIGURE 6.25 Welding of a copper elbow for the Prada store in Tokyo.

They may darken welds and make them less visible, but the point of statuary finishes and patinas is to react with the metal to create a particular color, and the metal at the weld is different.

Brasses with a low zinc content (C21000, C22000, and C23000) can be welded. C26000 is more difficult to weld and its color will change. This alloy has 30% zinc. Brasses with a low zinc content are single-phase alloys and tests should be performed when welding to determine if the color at the weld line is acceptable. GMAW and GTAW are best for welding these alloys.

Alloy C28000 is a two-phase alloy containing as much as 40% zinc. This creates difficulties in welding and the arc welding processes for the single-phase alloys works marginally on the C28000 alloy. The biggest challenge is the color. It will change at the weld as the zinc vaporizes and the metal cools.

The alloys containing lead, such as Architectural Bronze (alloy C38500), will not weld. They can be brazed, but welding the leaded alloys is not practical. The cast brass alloys are difficult to weld due to the inconsistent makeup of the casting. They are not as consistent as their wrought cousins. Many cast brass alloys include other elements, and some have small amounts of lead. These leaded castings will not weld.

Welding Silicon–Bronze Alloys

The silicon–bronze alloy C87300 is the cast alloy commonly used in art sculpture. This alloy has excellent welding characteristics. It has low thermal conductivity as compared to other copper alloys and the slag induced during welding provides protection from oxygen as the weld cools. The preferred methods of welding are gas shielded arc welding and gas tungsten arc welding, (GTAW). Alloy C87300 and the other silicon–bronze alloys can be welded with oxy-fuel gas and shielded metal arc welding (SMAW).

Welding the cast alloy C87300 can achieve a close match between the surface and the joint, but it will not be exact due to the variations in the cast alloy. Grinding and finishing the weld back followed by patination can conceal the weld joint, but if weathering is allowed to occur on the surface, the weld can become visible as in the two cast bronze artworks in Figure 6.26.

The left-hand image in the figure reveals a weld beginning to show as weathering has reached the patina. The right-hand image shows a porous weld in a casting.

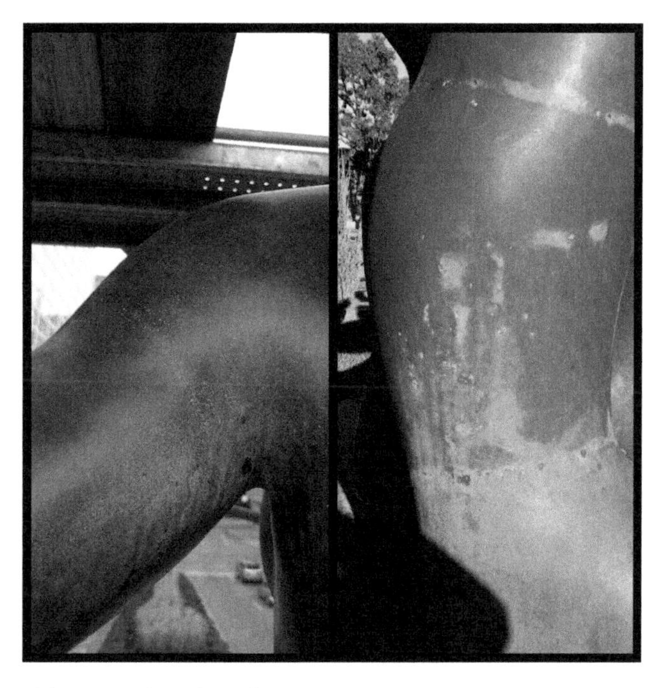

FIGURE 6.26 Visible welds on weathered cast bronze.

The parts being welded need to be clean and free of moisture and oils. When welding cast silicon bronze, be sure all sand and core material is cleaned from the surfaces. Preheating will drive the moisture away from the surface.

Welding Nickel–Silver, Phosphor–Bronze, and Aluminum–Bronze Alloys

Nickel–silver alloys can be welded. The zinc content creates issues similar to those of the brass alloys. The zinc vaporizes at lower temperatures than the nickel or copper. The biggest issue in welding nickel–silver alloys is the color match at the weld. It is not really possible to achieve a color match, and due to the esthetic quality of this alloy, an inferior color match can be a significant issue to overcome. Good design and reducing or concealing the welds or fusing the edges using GTAW techniques will work on thinner sections welded at edges. Nickel–silver alloys can be fusion stud welded effectively due to their lower conductivity.

Phosphor–bronze alloys can be welded using GMAW techniques coupled with preheating and sustaining the heat during the weld process. Other methods of welding (GTAW and SMAW) can also be used, but more selectively. The phosphor bronzes that are copper–tin alloys have a wide freezing range, meaning that they can crack along the weld as they cool from the welding operation.

Aluminum–bronze alloys used in art and architecture are usually not welded as they tend to crack. Alloy C61300 is preferred for welding situations, and even though it has slightly less than 7% tin, the addition of tin helps in the welding process. All of the arc welding processes can be used. A suitable shielding gas of argon or argon and helium should be used in most welding applications. Preheating is not needed for aluminum–bronze welding.

See Table 6.6 for a list of welding processes and their effectiveness when used on various copper alloys.

TABLE 6.6 Welding processes used on copper alloys.

Welding process	Performance	Description
GTAW	Excellent	Argon, helium, or mix
PAW	Excellent	Argon, helium, or mix
GMAW	Excellent	Argon, helium, or mix
SMAW	Good	Noncritical welds
Oxy-fuel	Good	Less warpage
Laser welding	Poor	Susceptible to cracking
Electron beam	Excellent	All alloys
Spot welding	Poor	Poor resistance at weld
Fusion stud welding	Good	Brass studs to copper

GTAW: gas tungsten arc welding; PAW: plasma arc welding; GMAW: gas metal arc welding; SMAW: shielded metal arc welding.

CASTING

The Lost-Wax Technique (Cire Perdue)

Casting with copper using the lost-wax technique is nearly as old as the metal's discovery and use by humans. The first copper was hammered from native forms of large nuggets of the metal, but not long after man found the metal could be melted and reshaped into new forms. This surely was a remarkable discovery. Imagine when the first broken copper tool or dish was melted down and poured anew into a wood or clay mold and a new shape was born. No other material would have offered this utility. Among the discoveries of man, this has to stand as one of the most remarkable. Now tools could be fashioned and shaped for a particular use, and hardened and sharpened into useful tools. When no longer useful or when damaged they could be remelted and refashioned. No stone, wood, or clay material known up to this time in history afforded this incredible versatility.

After weapons and tools, ornamentation was next. Early man found that he could shape clays, heat them to harden them, and pour molten metal into the clay forms. This was most likely the beginning of casting a one-sided detailed surface or ornament. At some point the discovery of how to form the clay over another material was made, and a similar process is still in existence today. This process is known variously as lost-wax casting, cire perdue, and investment casting.

As discussed previously, the process involves carving a form or shape out of wax followed by fashioning clay on the wax to make an outer refractory shell around the wax and capturing the detail carved into the wax. The mold is heated in an oven and the wax melts out, leaving a refractory clay negative of the original wax. Metal is poured into the mold cavity, and as it hardens it takes on the original wax form.

This process works for small objects, but for larger objects the technique is somewhat different.

A clay model is created and all the detail is put into the clay model. Another mold is made from this clay model. The mold is usually refractory sand. When complete, its contours match the original clay model contours. Hot wax is coated onto the surface to a thickness that will eventually be the thickness of the final copper alloy casting.

This wax mold is then filled with more refractory material. This is called a core. Wax tubes are created and located by skilled foundry workers. The wax tubes will provide the passageway for the metal to flow and vents for the air to escape as the metal flows into the mold. Refractory sand is packed around the form with wax vents and ducts in place. The entire object is placed in an oven and the wax melts out. Usually metal pins are set as well to hold the core in place once the wax melts out.

Molten metal is poured into the ducting system by way of a feeder system of reservoirs, and the molten metal flows into the cavity left by the molten metal. Once cooled, the metal casting is removed from the mold and the core of refractory material is removed. The pins are then cut from the surface and the vents and ducts are removed.

At the foundry, the mold is prepared to receive the molten metal. The mold is generally made of sand, but there are also ceramic molds, lost-wax investment molds, and several others. But sand molds are the most common in use for sculptures in copper alloy. Several examples of sand cast objects are shown in Figure 6.27.

FIGURE 6.27 Various examples of sand-cast shapes, from simple to intricate.

Difficulties in the casting process that manifest during and after the pour of the molten metals will translate to the finished product, sometimes causing the product to be rejected; the product must then be recast at the foundry. More often the cast is reworked to salvage what would otherwise be a good cast.

A few of the casting difficulties and their causes are listed in Table 6.7. These are nearly always addressed at the foundry before the sculpture or casting is delivered.

TABLE 6.7 Various surface maladies seen on castings.

Problem	Cause
Cracks	The metal pulls itself apart while it is cooling in the mold. This is often caused by imbalanced feed rates to different volumes of the sculpture. It can also be caused by a casting temperature that is too high or too low.
Cold shut	The metal does not join together where two or more streams feed the casting. A weak area or void is created. This is due to inadequate gating or dross entering the sprue. Wet sand or insufficient venting can also cause this defect.
Wormy surface	These are depressions near the gate. They appear as stretch marks or worm tracks. These are caused by inadequate gating, a high pour temperature, or low permeability of the sand used.
Sand wash	This is a rough surface, either lumpy or pitted, that is due to weak sand. Weak sand is often caused by an improper pattern removal from the mold or insufficient ramming of the mold.
Scab	This is a rough, slightly raised area on a casting. It can be caused by ramming sand that is too hard or too fine.
Core blow	This is a smooth depression on the inside of a cored casting and is due to a gas pocket above the core. It is caused by insufficient curing of the core or inadequate venting of the core.
Burning into sand	This is a rough sandy surface on the finish casting and is often caused by sand that is too coarse or too dry. Also, if the temperature of the pour is too high, this can create a rough surface, as the sand dries out too rapidly.
Gas holes	These are holes under the surface or at the surface of the finish casting. They can result from contamination in the pour or from the sand making up the mold being too wet.
Tin or lead sweat	This is whitish spots or a thin layer of whitish metal streaks on the surface. It is caused by inadequate venting or by interruptions in the pour process. It can also be caused by too small a gate, which will restrict metal flow and allow the metal to cool.
Rough or pitted surface	A rough surface on the casting or deep pits on the surface. This is due to inadequate care in the mold process and using a sand that is too coarse or allowing dross to enter the pour.
Solid inclusions	These are pieces of nonmetal materials in the casting from improper mold cleaning or an improper sand mixture.
Weak structure	This manifests as very small voids between the large crystals in the interior of the casting. It is due to inadequate gating and inadequate feed of the molten metal into the mold. It also could occur when gases are trapped in the melt.
Burning into core	This is characterized by roughness in the inner surface of the casting. This can be caused by a high temperature of the molten metal or from inadequate curing of the core.

Post-casting processes include removal of the core. The core creates the hollow interior space during the casting process. After casting the core is removed from the interior, leaving the casting hollow. There are instances where the core is not fully removed: interior sculpture will not be affected but outdoor sculpture with the core still in the cast will eventually have significant problems. All castings have some level of porosity. If significant amounts of the core are present inside the casting they will start to deteriorate and leach out onto the surface. They will appear as whitish blooms growing out from a central pore. Chapter 8 discusses this event, and Figure 8.30 shows an image of this defect.

Spinning

Spinning is a fabrication process used on wrought sheet material. Heat can be added to soften the metal and for performing shaping of thicker copper alloys. Spinning involves application of force to a spinning disk or plate of material. The material is stretched and shaped into a spinning mold as the force is applied. The greater the elongation an alloy and temper will allow under tension, the more able it is to be shaped via spinning. It is a characteristic referred to as the plastic–strain ratio.

The part produced is always symmetrical around a vertical axis. Hemispheres, cones, tubes, and dish shapes can be created. The conical shape of musical instruments is an example of spun copper alloys. Light sconces, such as the one shown in Figure 6.28, are another example of products that can be produced with this method.

Copper (C11000) is the simplest to spin. Usually it can be formed without interstage heating to anneal the metal for further shaping. All of the single-phase brasses can be spun, but those with higher zinc content may require annealing steps. Alloy C28000, Muntz Metal, and its sister alloy Naval Brass, C46400, cannot be spun. The phosphor bronzes, aluminum bronzes, and silicon

FIGURE 6.28 Spun and polished copper hemisphere light sconce.

bronzes are difficult to spin into anything but shallow disks because of their stiffness and cold working behavior.

Once the copper alloy parts are spun, they can receive finishing, polishing, statuary finishes, and patinas. Depending on the quality of the tooling and the skill of the operation, there may be radial lines imparted onto the surface. The lines are small and usually irregularly spaced. These lines are more pronounced on the side where the tool is brought to the surface.

Deep Drawing

As with spinning, the ability of a material to elongate while under tension determines its ability to be worked with deep drawing. This is the plastic–strain ratio referenced earlier: a measure of a material's ability to change in width and thickness as a tension load is applied. Many of the copper alloys have high, plastic–strain ratios. They can withstand stretching to a great extent before fracturing. Figure 6.29 shows a graph of a typical stress–strain relationship and the area where the metal will plastically deform until it fractures.

This ability is highly influenced by the mechanical behavior of the metal. Different copper alloys will work harden at different rates and this will affect the ability of the metal to undergo plastic deformation. Copper and many of the alloys of copper will elongate when subjected to a load. This allows for deep drawing of copper alloy parts. The alloys with the greatest percentage elongation have the ability to be deep drawn. Table 6.8 shows the elongation percentages for C11000 (commercially pure copper) under various tempers.

Temper induces stress into the copper by means of cold working operations. Extra spring has already moved up the curve to a point where very little additional forming can induce a fracture.

The other alloys have similar temper restrictions that, along with other mechanical properties, will determine the level of deep drawing they can undertake. Some of the alloys work harden rapidly and will have short and steep plastic deformation curves.

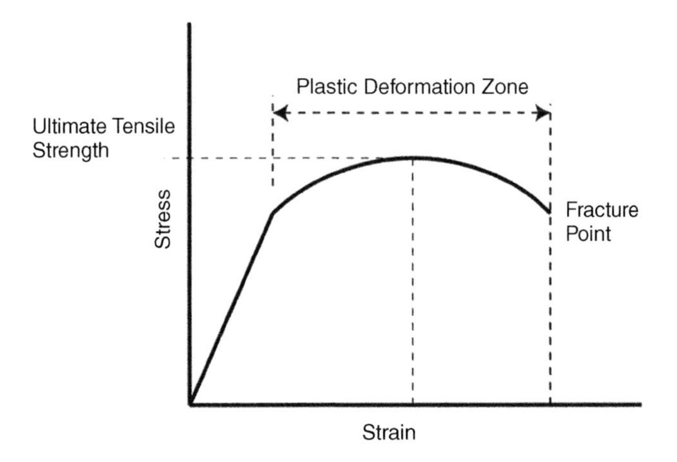

FIGURE 6.29 Plastic deformation zone before fracturing.

TABLE 6.8 Elongation of C11000 by temper designation.

Temper	Elongation %
M20 Hot rolled temper	50
H00 1/8 hard temper	40
H01 1/4 hard temper	35
H02 1/2 hard temper	14
H04 Full hard temper	12
H10 Extra spring temper	4

Stretch Forming

Stretch forming is a method used to form curved or contoured parts as opposed to a straight linear break form. The process involves gripping the metal sheet, fabricated form, or bar or tube along the edge and pressing it over a die while pressure is being applied. The metal is simultaneously pulled while being forced over a die.

Once the pressure is released, the copper alloy form takes the shape of the contoured die. There will be a slight necking of the material, and for polished surfaces some "orange peel" texture may be apparent as the grains of the metal realign under the force.

Similar to spinning and deep drawing, the copper alloys with the greatest elongation percentage, will stretch form with less difficulty. The annealed tempers are better suited for stretch forming and usually possess the greatest elongation for a given alloy of copper.

Custom Embossing and Custom Perforating

The ductility of copper alloys allows for deep embossing and perforating without creating a significant amount of residual stress. When metals sheets are shaped and punched to create perforations, often residual stresses build up around the initial indentation as the metal first elongates. In perforating and embossing, the metal is restricted around the zone where the work is to occur. This happens by means of a tool that clamps the metal surface or through pressure from metal-to-metal contact. The metal is restrained top and bottom. Where the metal is not restrained or supported, it elongates from the pressure into a cavity. In the event of embossing the pressure is limited. In the event of perforating, the pressure is increased to the point that a fracture develops as the slug of metal is sheared. In a real sense this is modern repoussé by means of computer-controlled mechanisms. The top and bottom images in Figure 6.30 show custom perforating and the middle image shows custom embossing on copper sheet.

Custom embossed copper was used extensively by Herzog & de Meuron when the firm designed the de Young museum of art in San Francisco (Figure 6.31). For this project it was necessary for

FIGURE 6.30 Custom perforating and embossing.

the fabricator to adjust the temper of the copper sheet to allow for the severe shaping that happens when numerous, discrete deformations occur. The metalwork hardens on a localized basis and this induced internal stress shapes the panel in unpredictable ways. To compensate, the fabricator adjusted the temper to allow for the added stress buildup.

Other means use rolls made of hardened steel with matching engraved designs. The metal is fed between the rolls and the pattern is pressed into the metal. A pattern can also be imparted by means of a cavity and hard rubber or by means of a cavity and hydraulic pressure, as shown in Figure 6.32.

FIGURE 6.31 Custom embossed copper at the de Young museum of art
Source: designed by Herzog & de Meuron.

To produce economical surfaces, patterns are engraved into matching rolls, of which one has a positive pattern and the other has a negative copy. The rolls are aligned and the metal is fed through in coil form or sheet by sheet.

All copper alloys in wrought sheet form can be patterned in this way. There are limitations on thickness for each process as well as pattern limitations. In custom embossing, the limitations are thickness (which usually needs to be less than 4 mm) and geometry. The embossed deformation must be outside the clearance of the hold-down mechanism used to constrain the sheet. Figure 6.33 shows the way in which the hold-down operates, and the minimum clearance needed. If the clearance is encroached upon, the hold-down will crush the previous embossed form.

This constraint doesn't exist for perforating, since there is no embossing to be crushed. However, there are constraints on the relationship between the size of hole and the thickness of the copper alloy sheet. The softness of many of the copper alloys is not as limiting as other, harder metals. Still, there are rules around the metal thickness and minimum hole diameter when using a piercing tool

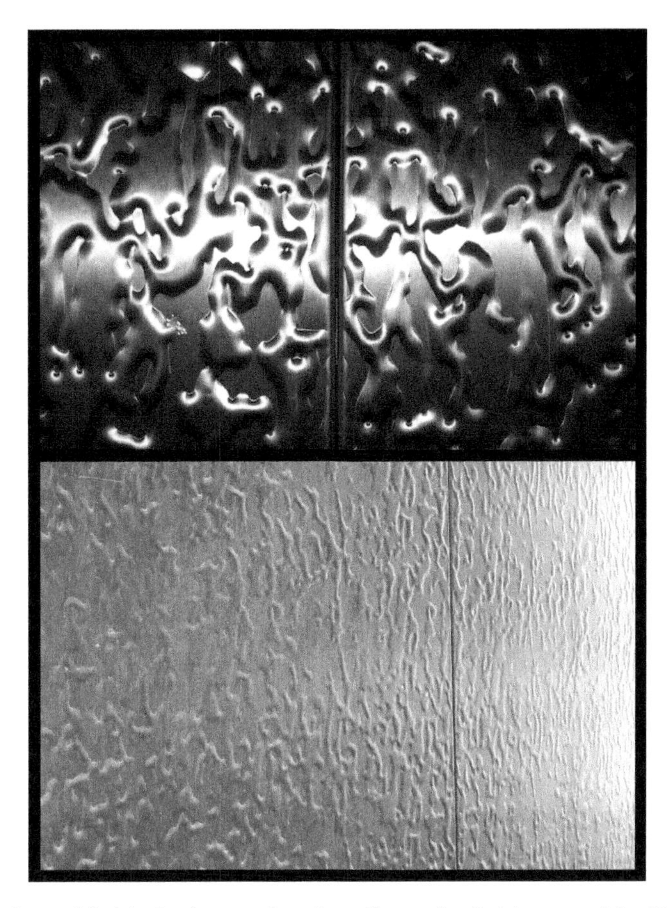

FIGURE 6.32 Roll embossed finish: (*top*) unaged surface; (*bottom*) a finish exposed for 12 years.

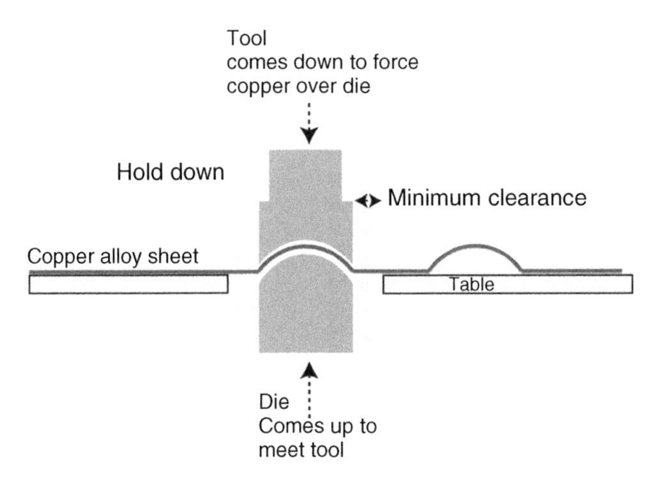

FIGURE 6.33 Diagram showing hold-down clearance needed for custom embossing.

to produce the hole. When the hole is very small the tool used to pierce the metal drags some of the metal around the edge with it, and as it pulls out of the hole this can drag on the tool and distort the edge. The harder the alloy the better the pierce is. For soft copper alloys, the edge-to-edge distance of the hole should be 50% or greater than the metal thickness.

These forming operations relate back to processes of shaping copper alloys thousands of years ago. Copper is the perfect metal for forming and shaping. Its plastic behavior is exceptional and has been used by artists and designers to accomplish some of the most remarkable work ever created using metal.

CHAPTER 7

Corrosion Characteristics

The color that forms on copper is evidence of time passing.

GENERAL INFORMATION

Copper and copper alloys are some of the most corrosion resistant metals in common use. The corrosion resistance of copper and many of the copper alloys approaches that of the more-noble metals in normal atmospheric exposures. Copper alloys are well suited for use in industrial, marine, and urban environments, where most other metals would deteriorate to the point of losing their utility. Copper and its alloys perform well in both fresh and salt waters. They are not indestructible, but they have performed over centuries of demonstrable service in harsh environments as well as in urban and rural exposures.

Examples of the long-term performance of copper alloys in our built environment are commonplace around the world. Christ Church in Philadelphia is an example in the United States (Figure 7.1). Its roof, which was clad in copper sheets imported from England in the late 1700s, is still performing remarkably well today. Inspections performed showed the rate of corrosion somewhere around 0.0008 mm per year.

In Europe there are numerous examples of copper roof and cornice work that date back centuries. Copper sheet was produced in various regions of Europe, and copper roofing was a significant surfacing material for many churches and cathedrals. Copper was also used as a roofing material in Asia, where temples and shrines often were clad in thin copper sheeting. In Hildesheim,

FIGURE 7.1 Christ Church in Philadelphia, Pennsylvania, has the oldest copper roof in the United States.

Germany, the cathedral roof is clad in copper sheets that date back to the end of the thirteenth century (Figure 7.2).

These surfaces, exposed to centuries of wear and pollution, are evidence of the long-term viability of copper and its alloys. Maintenance has been minimal: natural weathering coupled with good design details have enabled these structures to stand the test of time.

When copper alloys are exposed to the atmosphere you can expect the formation of a patina on the surface. The patina color and composition will vary depending on the alloying constituents and the exposure—whether urban, rural, or seacoast. The rate of the patina development will depend on the amount of moisture (in the form of condensation) present.

As corrosion takes place on a metal's surface, a decrease in free energy occurs. For the most part, the natural exposure of copper alloys to the atmosphere will begin a process of corrosion very slowly. There is an incubation period as oxygen and ionic diffusion occurs at the surface interface of the metal to the surrounding atmosphere. As the oxide grows, the free energy decreases and the rate of corrosion slows way down.

FIGURE 7.2 *Top*: Hildesheim Cathedral has the oldest existing copper roof in Europe. *Bottom*: Shinto Shrine in Japan, clad in copper.

Amazingly, measurements taken from specimens of functional, 300-year-old copper roofs in Europe show less than a hundred micrometers of thickness loss.[1] The patina that developed on these surfaces indicates the extent of the corrosion process. Once the patina is formed the rate of corrosion drops way down, as the surface becomes inert and mineralized into a permanent protective layer. In most urban settings this mineral is brochantite, a stable mineral form of copper,

[1]L. D. Fitzgerald, "Protection of Copper Metals from Atmospheric Corrosion," in *Atmospheric Factors Affecting the Corrosion of Engineering Metals*, ed. S. Coburn (West Conshohocken, PA: ASTM International, 1978), 152–159.

TABLE 7.1 Common mineral forms of copper.

Environment	Mineral	Formula
Rural	Gerhardtite	$Cu_2(NO_3)(OH)^3$
Urban	Brochantite	$CuSO_4 \cdot 3Cu(OH)^2$
Coastal	Atacamite	$CuCl_2 \cdot 3Cu(OH)^2$
Coastal	Nantokite	$CuCl$
Coastal	Eriochalcite	$CuCl_2$

whereas in coastal areas the minerals that develop incorporate chlorine and form atacamite, nantokite, or eriochalcite (Table 7.1).

Over the years of exposure these minerals form on the surface as the copper combines with pollutants to form beautiful patinas that have the added feature of being chemically inert. The Statue of Liberty (Figure 7.3) has developed brochantite over the surface of much of the copper. In addition to the brochantite, atacamite is also prevalent due to the coastal exposure. Copper alloys were used for the spikes on the headdress of Libertas because of the need for stiffness.[2]

FIGURE 7.3 Statue of Liberty.

[2]The outer edges of the spikes are made from a bronze alloy and the top and bottom surfaces are brass. S. Hayden, T. Despont, and N. Post, *Restoring the Statue of Liberty* (New York: McGraw-Hill, 1986).

The corrosion of copper and copper alloys is a change occurring at the surface of the metal. Depending on the exposure of the surface over time, different minerals may be evident on the same surface. Given the array of climate conditions, both natural and induced by humans, the variety of minerals that can form on the surface is vast. These can take on an appearance that is subtle and localized over the surface. This is the beauty of exterior copper surfaces. They do not look like paint but have a specific character afforded to them by time, moisture, sulfur, chlorine, and any number of other compounds that find their way to the surface.

When there is little sulfur or chlorine in the atmosphere, the surface of the copper will darken and the characteristic green will not develop. The rate of corrosion of the copper surface is dependent to a great extent on the layer of moisture on the metal surface and the length of time this moisture is present. Dry, arid conditions will have a slowing effect on changes in the copper alloy surface.

Copper has several corrosion products and each has a mineral equivalent. As copper corrosion occurs in various exposures, the expected path is toward the mineral compounds found naturally. There are other compounds that can develop when copper is exposed to certain pollutants, but otherwise compounds that are found naturally are normally developed. Copper wants to join with oxygen, and that thermodynamic drive is difficult to prevent.

Each mineral has a distinctive color. Table 7.2 lists a few of the more than 200 minerals that form with copper. In Chapter 1, Figure 1.11 shows several of the common ores of copper. As the compounds grow out from the copper alloy surface, the influence of the environment drives the formation of one type of mineral form or another. On a surface, there can be several mineral formations with subtle color differences. This is the beauty of the metal and the patina that develops on the surface.

The copper component of the copper alloys can be oxidized into two different states. One state is referred to chemically as copper (I) oxide. This has the chemical makeup of two copper atoms

TABLE 7.2 Various mineral forms of copper.

Corrosion compound	Formula	Mineral names	Color
Oxides	Cu_2O, CuO	Cuprite, tenerite	Red, black
Hydroxide	$Cu(OH)_2$	Spertinite	Blue to blue green
Carbonates	$CuCO_3 \cdot Cu(OH)_2$, $Cu_3(CO_3)_2(OH)_2$	Malachite, azurite	Deep green, blue
Sulfates	$CuSO_4 \cdot 5H_2O$, $Cu_4SO_4(OH)_6$	Chalcanthite, brochantite	Translucent blue, deep green
Chlorides	$CuCl$, $Cu_2(OH)_3Cl$	Nantokite, atacamite	Light green, dark green
Sulfides	Cu_2S, CuS, $CuFeS_2$	Chalcocite, covellite, chalcopyrite	Dark gray, blue gray, golden green
Phosphates	$Cu_2(PO_4)(OH)$, $Cu_2(PO_4)(OH)_3$	Libethenite, Cornetite	Olive green, blue
Nitrates	$Cu_2(NO_3)(OH)_3$	Rouaite, gerhardtite	Bluish green

FIGURE 7.4 Cuprous oxide on a door made of copper plate.

for each oxygen atom and is written as Cu_2O. The copper (I) oxide is known by the chemical name "cuprous oxide." This oxide normally produces a reddish tint, but it can also be yellow, orange, or brown depending on the thickness. This is a common state experienced by many copper alloys when exposed to humid environments. Cuprous oxide has formed on the surface of the copper door shown in Figure 7.4. The copper was uncoated and had a light Angel Hair finish, but after a short exposure in high relative humidity the oxide formed on the surface.

The other oxide that can form is copper (II) oxide, also known as "cupric oxide." This is similar to the mineral tenerite. It is black in color and will occasionally form on copper alloys, producing a dark, often blotchy surface (Figure 7.5). Copper (II) oxide often forms on top of copper (I) oxide as exposure to corrosive conditions persists. Cupric oxide is not as stable as cuprous oxide and will eventually form cuprous oxide as the surface is exposed further to oxygen and moisture.

Both oxides can be present on the surfaces of weathered copper and copper alloys. The cupric oxide, the darker of the two, can be removed with citric acid. Concentrated lemon juice, for example, contains citric acid and will remove the darker oxide. Removing cuprous oxide requires more aggressive acids, such as phosphoric acid. When both oxides are present, treating the surface with concentrated citric acid will remove the cupric oxide and leave the cuprous oxide untouched. Depending on the end finish desired, it may not be necessary to remove the cuprous oxide unless the intention is to arrive at a natural, bright copper alloy surface. If the desire is to restore the surface back to a statuary finish with a darkened and highlighted surface, the cuprous oxide will provide a workable base.

In all exposures, the development of one of these oxides is first to occur. Patinas, if they develop, form over these oxides and can be mixed on the surface. Patinas develop as moisture, sulfur, chlorides, and nitrates find their way to the surface. They become absorbed, interact with

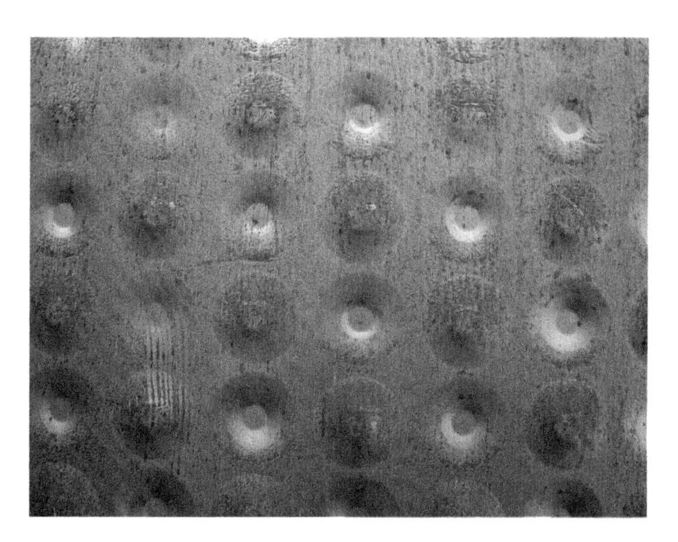

FIGURE 7.5 Black cupric oxide formation on copper exposed to low pH and humidity.

the oxide and hydroxide on the surface, and react with moisture and temperature to form hydrated copper compounds.

On highly polished, satin polished, and Mill finish surfaces, copper oxides will develop on the surface as a form of tarnish. In some cases only the oxide develops, and this is sufficient to protect the surface from further changes, as the oxide slows down further chemical interaction with the atmosphere.

When the protective coatings, lacquers, or waxes applied to the copper surface deteriorate and weather, exposing the base metal, the oxides develop at different rates and different intensities. Figure 7.6 shows images of three cast brass doors that have undergone different oxidation levels as the protective coating wore off and weathering took place on their surfaces.

Sculpture along the seaside is also prone to attack by chlorides. Once the protective wax or lacquer coating is breached, the chlorides can combine with the copper to form the minerals nantokite and atacamite. Telltale signs of this are green streaks and a pale green crust on the surface (Figure 7.7).

It must be noted that the patina that develops on various copper alloys will take different tones. They all start out in a similar fashion, developing a dark tone of cuprous oxide. Exposure to marine environments leads to the development of a streaky green patina, often with a gray-greenish color. Over time the color can turn brownish black, depending on the alloy. This progression applies to alloys exposed in urban environments as well. Dark tones and a splotchy surface will develop on the copper zinc alloys when left unprotected.

Salt deposits collect on surfaces near the seaside. These deposits go into solution when condensation forms on the cooler metal surface in the morning and evening. As the deposits go into solution on the surface of a patinated copper alloy, they interact with the copper and copper salt components on the surface. Figure 7.8 shows a prepatinated roof in Corpus Christi, Texas, a high-humidity seaside region. It is interesting to note that the sloping roof surface has aged differently from the vertical

FIGURE 7.6 Weathered copper alloy doors showing the presence of both oxides.

FIGURE 7.7 Casting made of the C87610 alloy near the seaside.

FIGURE 7.8 Prepatinated surface in Corpus Christi, Texas.

surface of the walls. Even the same panel shows a marked difference in color between the slope and the vertical. On the sloped region the color is changing, almost turning back to a dark tone. In reality, the patina is prematurely deteriorating. The chloride ion in the condensation is interacting with the copper. The prepatina is most likely a sulfur-based compound, and these often fail in chloride environments.

The vertical wall is slower to change because there is less condensation and fewer salt deposits here. The intriguing part is that the change in color is as if it were made that way. Eventually the sloped region will react with the chlorides and develop a natural patina that will be a different color, more blue green.

These corrosion products develop uniformly and pose little long-term concern for the copper alloys in a mechanical sense. The rate of corrosion diminishes after the first two years in all exposures, whether urban, rural, or marine.

CATEGORIES OF CORROSION

The alloying elements of copper alloys used in art and architecture will play an important role in determining the performance of the metal in a given exposure. Alloying elements added to copper can have a significant influence on the corrosion experience of the metal.

With the added influence of the esthetic conditions posed by art and architecture, even the most basic tarnish may be sufficient to impose appearance concerns. In polished surfaces, particularly those not protected, oxides will develop and the surface will become dull as people leave their fingerprints and accelerate the corrosion conditions. The decorative pole made of polished Cartridge Brass depicted in Figure 7.9 is showing tarnish where the public can touch the form.

The environment, the amount and nature of the moisture that collects on a surface, the temperature and humidity, and the pollutants on a surface will determine the risk of corrosion and the form this corrosion will take. Corrosion products that develop on a metal's surface are not necessarily

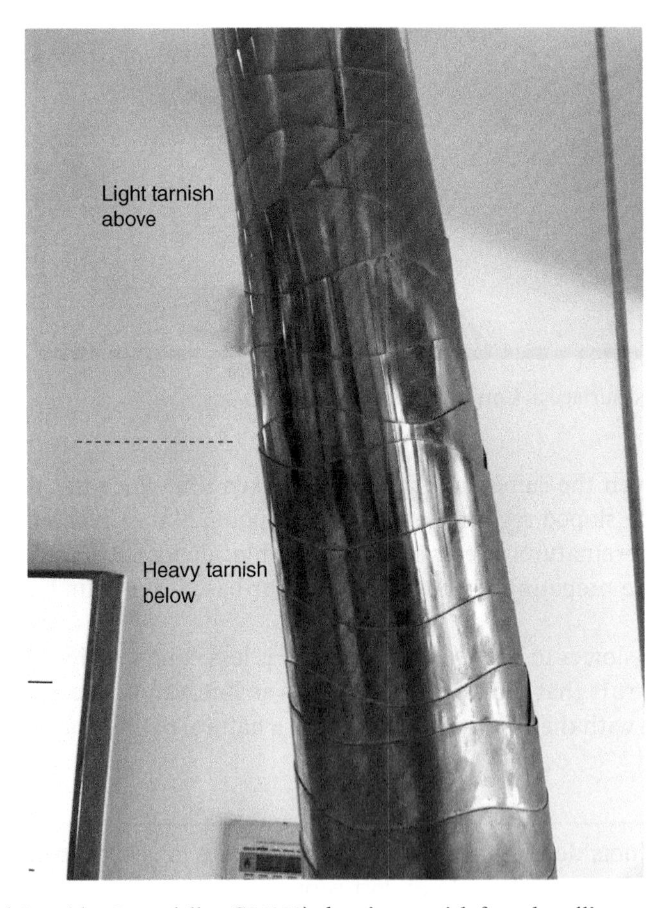

FIGURE 7.9 Polished Cartridge Brass (alloy C26000) showing tarnish from handling.

fixed in location. In actuality, variations in the amount of oxygen delivered to the surface can create areas that are more anodic in relation to those surface sites where exposure to oxygen is lower.

In vertical surfaces more moisture will be present in the lower regions as gravity pulls condensation to concentrate in these regions. When the surface is protected—say, by an overhang providing shade or protection from natural rains—the oxidation can increase, as more moisture remains longer.

There are variations in the homogeneity of the surface of copper alloys. Alloying elements will be present on the surface in the form of inclusions. Grain boundaries and areas where stress is more pronounced can alter rates of corrosion.

The frequency and form of cleaning the surface of the copper alloy will have a profound effect on the character of the corrosion. The removal of deicing salts, nitrates, chlorides, and sulfides from the surface will arrest the development of corrosion by removing substances that combine with the copper, as an electrolyte forms from available moisture, such as condensation.

Table 7.3 lists the various categories of corrosion and the likelihood of their occurrence for most art and architecture exposures. Each of these categories will be discussed in more detail in this chapter.

TABLE 7.3 Corrosion categories.

Type of corrosion	Risk	Identifying symptom
Uniform corrosion	Common	Tarnish darkening to an eventual patina of dark tones or green tones. Corrosion rate slows with time.
Galvanic corrosion, bimetallic corrosion	Common	Initial darkening and fuzzy gray-green matter at junction. Copper often is the more-noble metal, and thus the other metal in contact with it (in particular, aluminum and zinc) corrodes. Highly dependent on ratio of areas.
Pitting corrosion	Chloride exposure	Small, multicolored pits. Dark in center. Copper alloys are not as susceptible as other metals.
Dealloying, dezincification	Zinc alloys, aluminum alloys	Gray color in the center of a small spot on a copper alloy surface leading to a pit. Zinc content >30%. Additions of phosphorus or tin will thwart this corrosion.
Stress corrosion cracking	Corrosive atmospheres	Crack in the metal usually accompanied with a dark area of oxide. Zinc content >15%. Ammonia or sulfur dioxide atmospheres.
Fretting corrosion	Not common	Gouging or ripping of the surface.
Erosion corrosion	Not common	Constant rapid flow of moisture removes the oxide layer.
Intergranular corrosion	Not common	Attack along the grain boundaries.
Corrosion fatigue	Not common	Crack in the metal across the grain.
Crevice corrosion	Not common	Excessive pitting under seals or washers and laps.

Whether the copper alloy will experience one of the corrosion categories discussed in this chapter is dependent on five factors:

1. The driving force: electron flow
2. The ambient environment
3. The alloying elements
4. The regularity of cleaning
5. The soundness of the clear coating

The driving force behind corrosion of metals is the flow of electrons. This is a natural consequence of their temporary existence in metal form. This force is known as the "electromotive force." Each oxidation and reduction reaction a metal surface experiences has an electrical potential associated with it. If a chemical reaction is to occur spontaneously on the surface of a metal, there must be a driving force behind the movement of electrons. This can be created by an electrolyte that develops on the surface, proximity of another metal of dissimilar electrical potential, and variations in the surface of the metal itself that are created by oxygen imbalances or alloying imbalances.

The ambient environment considers the relative humidity the surface is exposed to, as well as pollutants, bird waste, chloride exposure, deicing salts, even handling and protective wraps. These will have an effect on the surface of copper and copper alloys, whether the finish is natural, statuary, or patinated.

The alloying elements, particularly with respect to the amount of zinc present, will contend with effects of dealloying, or the separation of the zinc and copper. Additions of tin or phosphorus can prevent dealloying of zinc-rich alloys, while other elements can afford a resistance to the corrosive influences of chlorides, and still others can slow the occurrence of visible surface tarnish.

For most of the metals used in art and architecture, the copper alloys included, the regularity and thoroughness of cleaning will enhance the ability of the metal surface to perform. Regular cleaning and the removal of substances such as sulfate deposits, chloride salts, and other reactive substances such as bird waste will enable the metal surface to perform in a consistent and predictable manner.

Clear barrier coatings are often the norm on copper alloys. Here, the expectation is for the oxide or patina to act as the final and constant appearance, and the intent of the design is to maintain the appearance of the metal in the condition in which it was first installed. The surface coating is an important barrier in maintaining the richness of the copper alloy finish, be it natural brass or bronze color, or induced to a colorful patina chemically. Without a sound barrier, the patina or induced oxide can be altered by environmental compounds that will either decay the oxide coating or interact with it in a way that is not esthetically pleasing.

Uniform Corrosion: The Initial Tarnish

Uniform corrosion can occur on all metals, but copper alloys tend to develop a thin oxide layer on the surface rapidly, and further corrosion will be slowed significantly unless other environmental influences are present. Copper alloy surfaces can develop various corrosion products depending on the exposure and environment.

The most common of the corrosion products that uniformly form on copper alloys is a substance called "tarnish." Tarnish is a thin oxide film that quickly develops on exposed copper and copper alloy surfaces. It closely resembles the mineral cuprite (Cu_2O). The copper on the surface seeks out oxygen and combines to form this common oxide known as cuprous oxide. Cuprous oxide has a reddish color and is initially semitransparent. Some copper hydroxide may also be present, but this would be in an intermediate development as the oxide transforms into the more stable cuprous oxide. As the oxide layer grows, this semitransparent film will generate interference effects, and colors ranging from yellow to purple may be visible. The film is virtually invisible until it reaches 40 nm. The clear oxide continues to grow and interference colors appear as the thickness exceeds 40 nm until it reaches a thickness of approximately 500 nm. These interference colors can be quite remarkable, or startling, depending on your perspective. Yellows, reds, blues, and purples may appear, and there may be a patchwork of color tones from element to element, or even across a single panel. This can occur with thin wrought copper sheets used as roofing or wall surface material on the exterior or interior. These surfaces are composed of sheets taken from the same coil or the same heat, yet the adjacent elements can form oxides at slightly different rates due to differences in moisture and oxygen. Exposure time plays a role as well. Copper alloys react with oxygen, and differences in the initial oxide development will be apparent depending on how a particular surface was exposed initially and for how long. These phenomena are short lived and will disappear as the oxide thickens and darkens into a rich brown color tone. If, however, a protective coating is applied, it will prevent oxygen from reaching the surface and these slight differences will remain.

The more moisture (in particular condensation) that is present, the more rapidly tarnish will form. On interior surfaces and in mild environments, the oxide grows slowly and is highly dependent on the surface and alloy. Human interaction in the form of handling and touching the surface will accelerate localized tarnish as finger and handprints appear.

Cupric hydroxide is often dispersed throughout the surface as the oxides grow. This converts to cuprous oxide as the surface heats up. The hydroxide form of copper is not stable and will quickly change to the more stable cuprous oxide (Table 7.4).

The oxide film develops by oxidation of the surface and by the diffusion of copper ions on the surface. At first the oxide grows outward, but as diffusion occurs this changes and the oxide grows inward, then stops as a barrier is developed. The metal surface seeks out thermodynamic equilibrium and a dissociation pressure develops, driving the copper ions to form an oxide (tarnish) on the surface. All metals have this relationship with air, the exception would be the noble metals, gold and platinum.

TABLE 7.4 Oxides of copper.

Copper oxide	Formula	Mineral form
Cuprous oxide	Cu_2O	Cuprite
Cupric oxide	CuO	Tenorite
Cupric hydroxide	$Cu(OH)^2$	Spertinite

The oxidation process takes place in stages, starting with a rapid absorption of oxygen at the metal-to-air interface. The initial bond is very weak. Next, dissociation of oxygen occurs at the interface, chemiabsorption of oxygen occurs, and a strong chemical bond develops. This begins to change the properties of the copper surface, reducing its reactivity with other substances. That is why forced patination processes require the removal of the oxide to increase the reactivity of the surface. Tarnish and mild oxidation on the surface inhibit the effectiveness of chemical processes used to patinate copper alloys.

The aluminum–bronze alloys (C61000, C61300, C61400, and C61500) have high tarnish resistance due to the aluminum component. The aluminum in these copper alloys develops a clear, aluminum-rich oxide on the surface, which resists further oxidation. The euro coin is made from a version of an aluminum–bronze alloy.

As the oxidation and diffusion of ions from the surface progresses, a passive film is developed. This film is impervious and very adherent. It forms a barrier to further diffusion of ions and the process of surface corrosion slows way down.

Transition to the Natural Patina

The cuprous oxide continues to develop on the surface of the copper alloy. As it grows outward from the surface copper ions migrate, and as the surface is exposed to other compounds, particularly sulfides and chlorides, copper compounds develop. When exposed to the atmosphere, chlorides and sulfides are the primary pollutants that combine with the copper on the alloy surface to form compounds equivalent to the minerals found naturally. This drive to combine with atmospheric elements essentially locks them into a thin layer of insoluble corrosion product that drastically slows the rate of decay. This very adherent, thin, colorful layer, provides copper and many of the copper alloys with the remarkable corrosion resistance characteristic that few metals can match. The process begins slowly and is not uniform across the surface, but instead localized in areas where ions are concentrated. Seams, edges, and lower regions of the surface are the first to react (Figure 7.10).

As the copper migrates to the surface when exposed to atmospheres containing sulfur or chlorine, a green film slowly develops. At first the green develops around the edges and on horizontal

FIGURE 7.10 Green developing along the lower edge of a copper panel.

surfaces where moisture is held on the copper surface. The moisture captures the pollutants and creates an electrolyte solution. Additionally, dew formed on the edges of the surface of the cold metal as water condenses runs down the surface and collects along the lower edges, effectively concentrating the pollutant compounds in a strong electrolyte along the edge. In the electrolyte, the chlorides and sulfides ions have a strong negative charge, which attracts the migrating positive copper ions and the various compounds of copper sulfide and copper chloride develop.

In seaside exposures, the most common mineral formed is atacamite, whose formula is $Cu_2(OH)_3Cl$. This copper chloride mineral form develops in areas near the sea and where high relative humidity exists. Other forms of copper chloride are cuprous and cupric chlorides, defined as the minerals nantokite and eriochalcite.

In urban regions, the surface of a copper alloy develops the brown cuprous oxide first as a film over the surface. As exposure to sulfides in the atmosphere collect on the surface, small crystals of copper sulfate begin to form in areas where dew or condensation has collected first. The surface darkens further, with the edges taking on a green tint from the growing crystals of sulfate. After about five to seven years, depending on the level of sulfur in the air, the surface will develop an overall transparent green tint that is heavier on the edges and lighter across the surface.

In some regions where the surfaces of copper roofs once developed a green patina in a decade or two, it now may take as much as 30–50 years to develop. This is because our air is cleaner today. The reduction of sulfur combustion particles due to clean air regulations requiring control of what is emitted into the air from the burning of coal, has lead to far less sulfur in our atmosphere.

Corrosion products that develop on the surface of copper are sometimes classified as their mineral representatives when they closely match the compounds that make up the copper minerals found in nature. Some of the mineral forms and their characteristics are listed in Table 7.5.

Surface patination is a form of degradation of the metal and causes a transformation of mechanical propertie. Patinas, for instance, lack flexibility and will crack under bending stress or impact. These compounds that form and create what we call a patina can be reactive and lead to the formation of other compounds or form into a very passive film: a noble film that protects the underlying metal against further corrosive attack. In almost all cases in art and architecture, the patinas that form eventually arrive at this protective noble film as equilibrium is achieved.

TABLE 7.5 A few of the minerals and associated colors of the patinas.

Type	Mineral name	Color
Oxide	Cuprite	Red
Oxide	Tenorite	Black
Oxide	Spertinite	Blue green
Carbonate	Malachite	Pale green
Carbonate	Azurite	Blue
Carbonate	Georgeite	Pale blue
Carbonate	Chalconatonite	Green blue
Carbonate	Rosasite	Blue green
Carbonate	Aurichaleite	Pale green
Carbonate	Claraite	Translucent blue
Chloride	Nantokite	Pale green
Chloride	Atacamite	Vitreous green
Sulfate	Chalcanthite	Deep blue
Sulfate	Brochantite	Green
Sulfate	Antlerite	Green

> The natural patina development on copper and copper alloys actually removes pollutants such as carbon dioxide, sulfur, nitrogen, and chlorine from the air and locks them permanently into the colorful patinas that adorn the roofs of some of the most interesting urban structures that have stood the test of time.

Galvanic Corrosion

Galvanic corrosion is a common corrosion concern with all metals. It is little understood by many in the art and architecture community. Table 7.6 lists the various metals in order of the electrical potential of two different metals exposed to seawater. Other charts use different exposures, but

TABLE 7.6 Electrical potential of various metals in flowing seawater.

	Scale	
Anodic polarity	**Voltage range**	
(more active) end of the scale: least noble metals	−1.06 to −1.67	Magnesium
	−1.00 to −1.07	Zinc
	−0.76 to −0.99	Aluminum alloys
	−0.58 to −0.71	Steel, iron, cast iron
	−0.35 to −0.57	S30400 stainless steels (active)
	−0.31 to −0.42	Aluminum bronze
	−0.31 to −0.41	Copper, brass
	−0.31 to −0.34	Tin
	−0.29 to −0.37	50/50 lead–tin solder
	−0.24 to −0.31	Nickel silver
	−0.17 to −0.27	Lead
	−0.09 to −0.15	Silver
	−0.05 to −0.13	S30400 stainless steels (passive)
	0.00 to −0.10	S31600 stainless steels (passive)
	0.04 to −0.12	Titanium
	0.20 to 0.07	Platinum
	0.20 to 0.07	Gold
	0.36 to 0.19	Graphite, carbon
Cathodic polarity end of the scale: more-noble metals		

seawater is often used due to the electrolytic behavior of the water. There are a number of factors that will determine whether a metal in a galvanic couple will corrode or not. The difference in the potential shown on the chart is one critical component. The properties of the electrolyte and the resistance of the circuit between the two metals (that is, how connected the two metals are) and second, how well will the current flow in this connection.

Many design professionals look at a metal's position on this chart and arrive at the conclusion that certain metals should not be joined. Whether corrosion will occur, how rapidly it will occur, and which metal will corrode requires an understanding of the principles behind this mysterious tendency of what are termed "dissimilar metals."

The values shown for each metal listed in Table 7.6 are derived from tests of pure metals in flowing seawater, since seawater is a sufficiently strong electrolyte. Water containing deicing salts could just as well be used, as it forms another strong electrolyte, common to many northern climates. The electrical potential is the energy per unit area as measured in volts. It is based on the fact that two different metals will have a different potential charge when an electrolyte is present. When the charge is sufficiently different and the metals are joined, the flow of free electrons occurs from the negative (anode) to the positive (cathode) through the electrolyte. Figure 7.11 shows a typical cell and the parts needed: two different metals, an electrolyte to carry the ionic charge, and a physical connection to complete the circuit.

A current develops and will flow from the anode, or more electrically negative metal, to the cathode and then back again through the metals where they are connected, thus completing the circuit. The anode side oxidizes in the solution as metal ions on the anode surface join with the oxygen and hydroxide in the solution. Certain metals, such as aluminum and titanium, rapidly develop

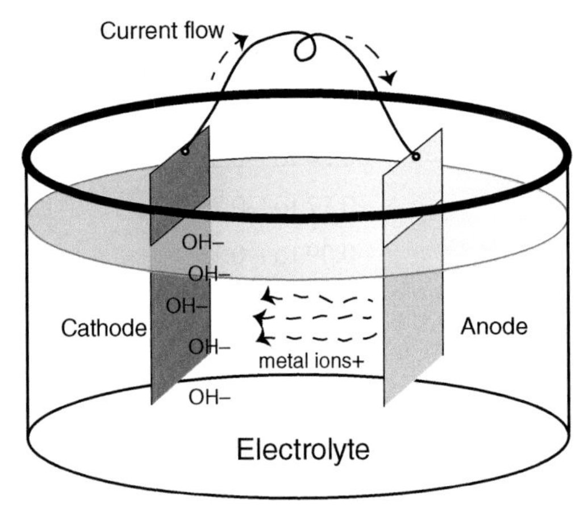

FIGURE 7.11 Makeup of a galvanic cell.

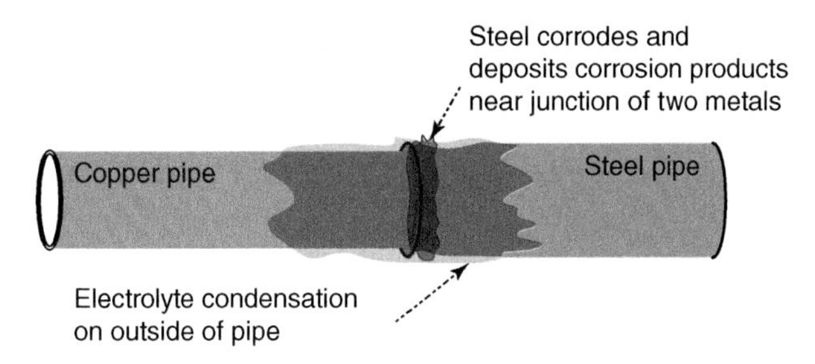

FIGURE 7.12 An idealized cold-water connection where a copper pipe is joined to a steel pipe. The steel will corrode.

an oxide layer that reduces the free energy available on the surface and thus can slow the rate of corrosion. The galvanic chart shown in Table 7.6 does not take this into consideration.

When a galvanic couple is developed, the anodic metal is said to be sacrificial to the cathodic metal and often prevents the cathode from corroding. So in the instance of a steel pipe being joined to a copper pipe with an electrolyte present, the steel will corrode and deposit iron oxide around the joint, as depicted in Figure 7.12. Little-to-no corrosion will occur on the copper pipe, while the steel pipe will undergo significant and rapid corrosion as long as the electrolyte is present and electrons are allowed to flow through the electrolyte. Note that condensation is the vehicle of ion flow. If the pipe is dry the rate of corrosion will be very slow, if there is any at all.

There are two processes involved in galvanic corrosion; the first one is the corroding or dissolution of the anodic metal. In the example above this process is acting on the steel pipe and the reaction is:

$$Fe \rightarrow Fe^{2+} + 2e$$

Here the iron ions are dissolved into the electrolyte.

The second process that happens is that the dissolved oxygen found in the electrolyte goes through a process called "reduction." In reduction, electrons are added. In this case they are gained from the iron. This is also known as a "cathodic reaction."

The oxygen along with the water molecule and the electrons that are now released from the iron, react as follows:

$$O_2 + 2H_2O + 4e \rightarrow 4OH^{4-}$$

When this happens, the iron goes into the solution as a positive ion and combines with the negative hydroxide and forms iron hydroxide, or rust. This comes out of solution and deposits on the pipe.

This is referred to as an "oxidation–reduction reaction." This is the basis for galvanic corrosion. It is an electrochemical action. There is a flow of electricity through an electrolyte by means of ions in the solution. An electrochemical reaction is defined by the transfer of electrons in a chemical reaction known as "oxidation and reduction." The stripping of electrons from the iron is called "oxidation." The process is not necessarily associated with the oxygen. An anodic reaction is an oxidation reaction as electrons are stripped and now available.

By itself, the anodic metal may not react in a similar electrolyte when not coupled with the cathodic metal, but when coupled the corrosion process will continue as long as the electrolyte is present.

Additionally, the two metals do not necessarily need to be physically connected—they only need to be electrically connected. The most common galvanic or bimetal corrosion condition occurs when they are physically connected; however, if the two metals are joined by conductive media, galvanic corrosion can still occur.

For galvanic corrosion to occur there must be ion formation at the anode, in this case the steel pipe. Here oxidation is occurring and the iron is corroding. At the same time there must be an acceptance of the electrons at the cathodic site (the copper pipe).

In this case the corrosion occurs only at the site where the two metals connect. The entire steel pipe is not affected, only the area near the copper pipe, where the ion flow is the greatest.

Copper salts can also corrode metals. What happens in that case is the salts become deposited onto adjacent or nearby metal surfaces and these cause isolated galvanic cells to develop. Usually this is a stain that grows until the less-noble metal corrodes. You can see the green stain developing over the years on limestone or concrete surfaces just below exterior copper surfaces. This same oxide, when allowed to pass onto steel surfaces, galvanized steel surfaces, and zinc surfaces, will corrode these metals.

For example, a copper gutter installed over the top of a zinc roof can corrode and eventually perforate the zinc roof if the condensate that forms on the copper drips onto the zinc. The condensate captures small amounts of copper ions and these are deposited onto the zinc, only to generate numerous small corrosion cells. Similarly, condensation from a rooftop air handling unit, passing through a copper tube and running over a roof into a steel drain or galvanized steel downpipe, will cause small galvanic cells and lead to corrosion of the less-noble steel or galvanized steel.

Copper on the electromotive scale is more cathodic than aluminum, zinc, steel, and iron. The voltage potentials for copper are negative, but these other metals are more negative. The size of this difference is often thought to be the determining condition. Generally, a potential difference of 50 mV or more is considered a situation wise to avoid. But this is not the only factor.

Determining Factors in Galvanic Corrosion

What determines whether one metal or another will corrode and the rate of that corrosion when in bimetallic coupling is related to the amount of current flowing between the metals. The passive film that develops on metals such as titanium, aluminum, and stainless steel make them passive and current flow is reduced. However, for aluminum, copper, and copper salts can set up corrosion cells with the aluminum oxide.

The amount of current that flows between two dissimilar metals with different electropotentials is affected by temperature, electrolyte composition, oxide, and other protective barriers—and most significantly, by the ratio of areas. Additionally, oxygen is needed. Corrosion will only occur if there is dissolved oxygen in the moisture that is present. Natural moisture, exposed to the air, rapidly acquires oxygen.

Determining factors for galvanic corrosion:

- Temperature
- Electrolyte
- Ratio of areas

Temperature Effects

Corrosion is a electrochemical reaction that occurs more rapidly in warm regions than in cooler regions. Chemical reactions that occur when metals corrode are slowed down when the temperature is cooler. Bimetal corrosion speeds up in warm and humid conditions. Chemical reactions occur more readily when heat is added. For example, deicing salts are in common use in conditions of very intense cold. Chemical reactions slow way down. The deicing salts may do their job on melting ice, but corrosive activity on metal surfaces does not intensify until the temperatures rise sufficiently in the springtime.

Electrolyte Effects

An electrolyte itself has to have the capacity to transfer current through it. If the capacity is weak, the current will be less. Electrolytes with a high capacity are those that have an ionic makeup, such as acids and alkalis, in particular those that contain chloride ions. Seawater and condensation, particularly condensation where the use of deicing salts is prevalent, make for strong electrolytes. Condensation from dew forming on a metal surface is more of a concern than rainwater. Condensation forms slowly and remains on the surface longer. Deionized water and distilled water are poor electrolytes. The more ions in solution, the more conductive the solution is and therefore the stronger the electrolyte. Rainwater has a low capacity to carry a current unless or until the water comes in contact with pollutants. The pollutants can be airborne or on the surface of the metal and as the water sits on the surface, a more powerful electrolyte can form.

If the electrolyte is very thin, even in the case of strong electrolytes, the corrosion is will be limited and generally right at the point of contact where the two dissimilar metals touch. The corrosion may still be rapid at the point of contact, since the resistance to electrical flow is lowest there.

The amount of oxygen and the ability of the oxygen to diffuse to the surface of the metal is another consideration in galvanic corrosion. If the electrolyte is sufficiently aerated—that is, contains a lot of oxygen—the rate of galvanic corrosion will increase. For copper and copper alloys the amount of oxygen in the electrolyte can render them less noble.

Ratio of Areas

The other very important condition of galvanic corrosion is the area relationship of the cathode and anode. If the cathode (the more positive metal area[3]) is significantly greater than the anode metal,

[3]In a galvanic cell, the cathode is considered positive and the anode, the part giving up electrons is considered negative. In an electrolytic cell, the anode is considered the positive while the cathode is negative.

the rate of corrosion will accelerate. If the anode area is significantly greater than the cathode, the rate of galvanic corrosion of the anode will be lessened.

For example, if a stainless steel fastener or clip is used to fix a copper roofing panel with a cross-section significantly larger than the stainless steel, the area of stainless steel in relationship to the area of copper is very small. The galvanic current will be constant between the two metals, but the corrosion per unit of area of the copper will be low because the current density is low. The stainless steel fastener is needed for the strength it will provide. Stainless steel is more noble than the copper when in a passive state, and the area of copper is significant in relationship to the small area of the stainless steel fastener. The stainless steel will be protected by the copper anode in this case and corrosion of the stainless steel, even when exposed to a strong electrolyte, will be negligible. The copper will experience very little corrosion due to the larger area in relationship to the stainless steel.

The exception to this can occur when the stainless steel goes active. "Active" means that the stainless steel is corroding, as in a highly active salt or halogen environment. Figure 7.13 shows stainless steel support members under a large copper alloy surface. The surface was set in a pool

FIGURE 7.13 Stainless steel supports and the C22000 copper alloy in a chloride environment.

of water that used chlorine as an anti-algae measure. The chlorine content was sufficient to create fumes under the copper alloy surface, and this combination of chloride fumes, copper alloy, and copper salts corroded the stainless steel in a few months of exposure.

Chloride containing electrolytes will move the stainless steel to an active form, and this will switch the ratio of areas and cause the stainless steel to experience intensive corrosion.

For galvanic corrosion to occur, you need an electrolyte, a source for oxygen and some form of connection to allow electrons to flow from one metal to the next. Preventing any one of these and galvanic corrosion will be impeded. See Figure 7.14.

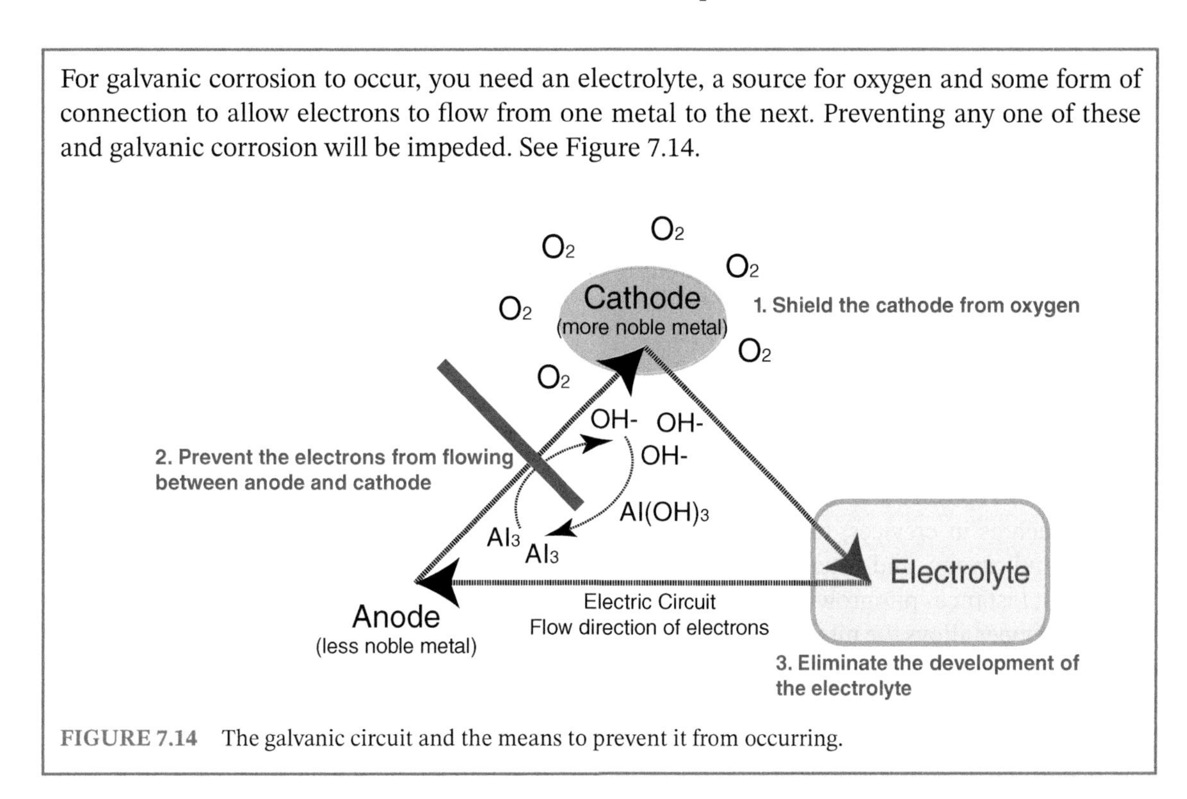

FIGURE 7.14 The galvanic circuit and the means to prevent it from occurring.

Brass and bronze alloys of copper used in art and architecture are normally coated with wax or lacquer, which will negate the development of corrosion products. If, however, they are attached to aluminum, steel, or zinc without a sound separation layer, they will cause the less-noble metal to deteriorate. It is important to coat the copper alloy sufficiently rather than the less-noble metal. The reason this is critical is, again, the ratio of areas. If the coating on the less-noble metal is scratched or breached in any way, a small area will be exposed to bimetallic corrosion through the coupling with the larger area of the more-noble metal. With copper and aluminum couples, anodized aluminum may offer some electrical resistance, but the porosity of the anodized aluminum will draw in the copper salts and hold them. As condensation develops, the copper salts will set up corrosion cells and the aluminum will be affected.

This is one of Faraday's laws: a given current will pass between the anode and cathode in a galvanic cell at a proportional rate. If, for example, the cathode area is 10 times larger than the anodic metal area, then the current will be 10 times as great passing through the anodic metal and corrosion of the anode will be rapid.

Steps in Preventing Galvanic or Bimetallic Corrosion

- Design the joint properly to prevent water from collecting.
- Use metals close together in the galvanic series (50 mV or less).
- Respect the ratio of areas. Avoid small areas of less-noble metals.
- Use a nonconductive coating to separate the metals.
- Coat the more-noble metal rather than the anodic metal.
- Prevent electrolyte from collecting at the junction of the two metals.
- Prevent copper salts from being deposited on less-noble metals.

Pitting Corrosion

Copper alloys in certain environments can develop pits, but copper alloys are less susceptible than other metals, such as stainless steel or aluminum. Pits can form where chlorides are prevalent, as with seawater exposure and deicing salts exposure. Localized pitting can occur where moisture is trapped, such as in crevices, where splash zones around seaside areas have a constant supply of chlorides and oxygen, and when deicing salts are left unchecked.

In most instances pits grow to a certain point, then slow down or stop as corrosion products seal the pit. In copper alloys the pits tend to go just so deep, then expand outward and leave a roughening area around the initial pit—unlike stainless steels, whose pits tend to go deep into the metal.

Ammonia and certain woods with ammonia biofouling can attack copper and copper alloys. If the metal is allowed to remain moist, severe pitting can develop.

The most pit resistant of the copper alloys are aluminum brasses. Copper–nickel alloys and low-zinc brasses are also relatively pit resistant, while silicon bronzes are more prone to pitting corrosion.

Bronze Disease

Bronze sculptures near the seaside can develop a pitting corrosion known as "bronze disease." Bronze disease is a condition that can also develop on copper alloy surfaces other than bronze alloys that are exposed to high-chloride environments.

The condition usually appears as a large pale green spot or patch in a localized area on the surface. Nearly all bronze sculpture near the sea can experience this corrosion condition. It is not reversible, and if allowed to continue can spread and damage the piece. Surface transformation does not require elevated temperatures for this corrosion type.

To stop it, you need to get as much of the chloride off of the surface as possible. The corrosion is thought to begin as a layer of cuprous chloride forms and copper below the layer dissolves and redeposits. This prevents oxygen from getting into the pit region, and thus the protective cuprous oxide cannot form. Mechanical means may be needed to remove the cuprous chloride. Use deionized water to thoroughly rinse the surface and remove the chloride ion. Apply a solution of sodium sesquicarbonate with a pH of 10 to the area. This will neutralize the acid character of the cuprous chloride. Sodium sesquicarbonate is a detergent-like substance and is safer than borax or trisodium phosphate as a cleaning solution. Rinse thoroughly and apply a wax or lacquer coating containing benzotriazole in an effort to protect the surface from chloride exposure.

Dealloying/Dezincification

The addition of zinc to copper results in many beneficial and esthetic characteristics. As zinc is added to copper, it becomes stronger and different color tones are exhibited. The color of the various alloys change from a reddish copper tone to a golden yellow. However, at just over 15% zinc content copper alloys are prone to a condition called "dealloying," also known as "dezincification."

Dezincification occurs when the alloy is subject to a corrosive environment, such as a high-chloride seawater exposure. The zinc component of the alloy begins to separate on the surface. When this occurs, silvery gray spots appear on the surface as the zinc is selectively dissolved from the alloy mix. In time the brass is replaced by weak, porous copper. Dezincification can also occur in protected areas where absorptive substances remain on the surface. Ammonium hydroxide and ammonia-based cleaning solutions left on the surface of brass alloys can lead to this form of corrosion. Several commercial brass cleaners use ammonia-based solutions to remove light tarnish and oxide from the surface. If these are not completely removed from the surface, the potential for developing a dealloying spot will increase. When remnants of these cleaners are allowed to remain on the surface in a confined area, the rich zinc–copper alloys can show signs of this corrosion.

In dezincification, one of two conditions occurs: the zinc can either be removed from the copper alloy, leaving behind a porous friable copper deposit, or both the zinc and copper can be dissolved and the copper redeposited on the edge of the pit. When the zinc is selectively removed, the process proceeds very slowly as the zinc atoms migrate to the surface. The process can occur at room temperature and where few, if any, chloride ions are present.

One common form of the corrosion often resembles pitting corrosion because a pit is formed as the corrosion takes place. The pit can be deep as the alloy dissolves. This form is referred to as "plug-type dealloying" and is the more common of the dealloying attacks on brasses used in art and architecture.

The dezincification process can be autocatalytic and continue even when the product is in storage. Figure 7.15 shows images of corroded alloy C26000, known commonly as Cartridge Brass. Some regions are more severely corroded, and there is a porous layer of reddish copper left behind on the surface. Other areas have whitish spots in the middle of the surface, while on the whole a thick tarnish of cuprous oxide covers the surface.

Dezincification corrosion is not reversible. The products causing the corrosion can be removed, but the condition has resulted in the removal of alloying elements and a small pit will remain. When the corrosion particles are removed, the surface will be slightly rougher. The surface shown

FIGURE 7.15 Dezincification on a C26000 surface.

in Figure 7.15 started out as a mirror-polished surface. It had been cleaned over the years, but the cleaning substances used may have contained ammonia; when the part was placed in storage, the surface reacted to the ammonia and trapped moisture. Figure 7.16 shows a microscopic view of the corroded regions. The scale is 0.1 mm.

For dezincification to occur, the alloy must have more than 15% zinc. The optimum environment for this type of corrosion is one rich in oxygen and carbon dioxide, with any moisture present being stagnant. The alpha-beta phase alloys, C28000 and C38500, can be more prone to dezincification. They have a high zinc content and the dual phase of the crystal structure lends itself to intergranular attack.

To prevent or inhibit the occurrence of dezincification, additions of tin in a proportion of as little as 1% will suffice. Naval Brass (alloy C46400) and manganese bronze (alloy C67500) both have tin added for this reason. The addition of tin to copper alloys improves corrosion resistance in marine environments, increases strength, and decreases the tendency for dezincification to occur.

The copper–aluminum alloys are also potentially subject to dealloying; in these alloys, it is the aluminum that is selectively removed. Alloys containing more than 8% aluminum can be prone to dealloying in severe chloride environments.

Stress Corrosion Cracking

This type of corrosion is an intergranular corrosion that manifests as a crack. To develop stress corrosion cracking, the copper alloy part must be subjected to stress and corrosive environments. The stress can be generated from applied forces or from thermal movement repeated over the

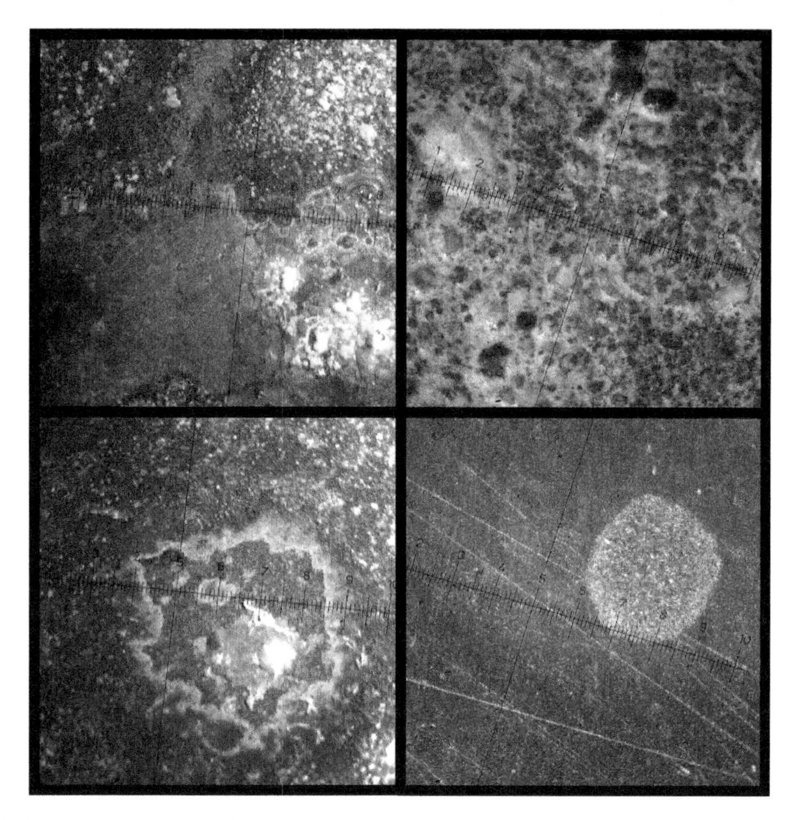

FIGURE 7.16 Microscopic images of dezincification corrosion.

service life. The corrosive environment usually consists of an ammonia-based exposure, such as exposure to cleaning fluids or wood preservatives. The crack will develop in the region where the stress is most significant. Propagation of the crack has little relationship to the grain size or alloying elements in the copper alloy. However, copper alloys with less than 15% zinc are less prone to stress corrosion cracking.

Copper roofing installed over some fire-resistant wood substrates, particularly those containing ammonia compounds, can develop a crack in the regions subject to stress.

Other corrosion types are identified in Table 7.3; however, for the most part, they are rarely applicable to the use of copper alloys in art and architecture. They are briefly discussed here but there should be limited concern about these corrosion types for art and architectural projects.

Fretting and Erosion Corrosion

Fretting and erosion are similar corrosion conditions that result from surfaces interacting. Fretting corrosion occurs when tiny surface asperities rub together on two adjacent surfaces. There are many double lock seam copper roofs, for example, that have been in use over the centuries.

No major fretting conditions develop, where the clips or seams rub against one another from thermal expansion and contraction. Properly installed, the metal surfaces slide over one another without developing fretting corrosion.

Some copper alloys are designed for bearing plates of bridge supports, large earthmoving equipment, and even airplane landing gear. Copper alloys have been used for centuries as a metal for hinging, linkages, gears, and other sliding and bearing surfaces. The tin bronzes, in particular those with additions of phosphorus, have excellent bearing strength and have a working affinity with steel. Sometimes called "bearing bronze," these alloys resist seizing up as other metals might and offer a self-seating characteristic when under load.

Erosion corrosion can also occur when water is constantly running over a surface and removes the oxide from the surface. This is rarely an issue in architecture or art projects. Fountains sometimes use a weir made of copper alloy. Water is constantly flowing over the weir surface. This may impede the oxide growth but its highly doubtful this will pose an erosion corrosion concern over the useful life of the fountain.

Crevice Corrosion

Crevice corrosion can occur when surfaces are bound tight to one another, but not so tightly that moisture is prevented from entering the gap between the metal surfaces. Crevice corrosion can occur where a gasket substance is separating two surfaces and the gasket absorbs moisture and holds it against the surface of the metal. The copper alloys used in art and architecture generally do not experience crevice corrosion. Yellow brass tubing can see a form of this corrosion when ends are wrapped or encased in protective wraps and moisture is present.

Intergranular Corrosion

Intergranular corrosion can occur in certain alloys exposed to hot steamy environments or high-sulfide environments. Rarely are these environments the purview of art pieces or architectural surfaces. Intergranular corrosion occurs along the grain boundaries of an alloy. The grain boundaries are electrochemically different from the balance of the metal and minute changes can occur. The aluminum bronzes are more prone to this corrosion than other copper alloys because Al_2Cu can precipitate at the boundaries and create cathodic regions where the balance of the metal is anodic.

Fatigue Corrosion

Fatigue corrosion is similar to stress corrosion cracking, with the latter being the more common. Fatigue corrosion, or corrosion fatigue as it is often known, is the loss of mechanical strength due to exposures to certain environments while under repeated loading and unloading. Fatigue corrosion can be experienced by copper alloys in environments rich in ammonia. There were periods in the past when wood backing was impregnated with ammonium salts for rot prevention. These salts would outgas and affect the copper or copper alloy cladding.

The decaying of some substances—urea-based materials, for example—can create ammonium-rich environments that can promote fatigue corrosion as the metal undergoes normal thermal expansion, cyclic wind loading, or other stress conditions. The condition is not a commonly experienced corrosion category by copper alloys used in art and architecture.

ENVIRONMENTAL EXPOSURES

The International Standards Organization (ISO) has defined five distinctive environmental exposure categories. These categories help to pinpoint which alloys are most suitable for a given exposure. The categories are listed in Table 7.7.

This classification of distinct regions helps to provide a framework for both the selection of the alloy and the approach to design and maintenance. As the category numbers listed in Table 7.7 get higher, the potential for exposure to various compounds increases, and with it the potential for more destructive forms of corrosion.

As a rule, the copper alloys perform very well in all of these exposures, but they do this by developing a surface layer of adherent chemical compounds that often lead to an esthetic decline.

The initial film that forms on copper alloys, regardless of exposure, is cuprous oxide (Cu_2O). This thin adherent oxide layer is what makes up the tarnish on interior exposures and what forms initially in arid environments. This oxide provides protection to the underlying metal. Depending on the environment, this film undergoes a natural transformation to sulfates or oxychlorides. Carbonates may also be present as this natural film grows.

Consider the first ISO category, C1. In these environments copper alloys will form a light oxide on the surface. Commonly referred to as tarnish, this light oxide will form slowly as the surface combines with oxygen to form a denser cuprous oxide. As the layer grows, interference colors may become apparent on the surface as light passes through this semitransparent oxide film.

If there are other substances on the surface (such as fingerprints, cleaning compounds, or the remnants of polishing compounds), the development of this oxide will be altered. Figure 7.17 shows

TABLE 7.7 ISO environmental exposure categories.

Category	Environmental description
C1	Indoor environment; arid desert environment
C2	Indoor environment, unheated; rural environment
C3	Outdoor environment, high humidity; northern coastal; clean urban environments
C4	Indoor environment, chemical exposure; polluted coastal; urban north
C5	Industrial; high humidity, high pollution; high salt-exposure environments

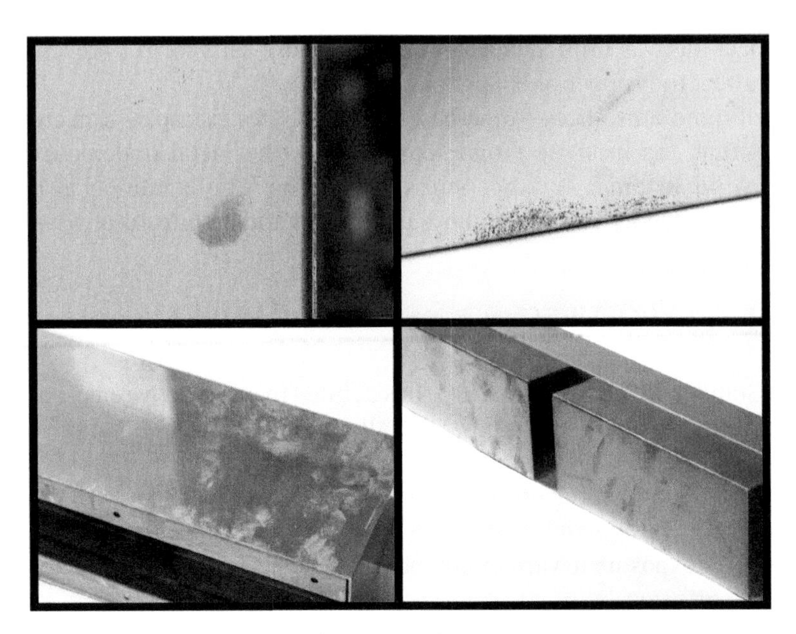

FIGURE 7.17 Fingerprints and handprints on different surfaces.

stains on polished copper and brass surfaces from light corrosion resulting from oils left behind from the handling of the metal surface. These oils have initiated the development of a deep tarnish. This light oxide can be esthetically displeasing and is removable with a commercial-grade copper or brass polish. Fingerprints are composed of organic oils and fats called "lipids," which are amino acids and water produced by the body. There are often salts intermixed in the oils, but the amount of chloride in a fingerprint is insignificant. The water will evaporate but the oils and fatty substances will remain. It is these oils and fats that are so persistent and difficult to move. Many cleaners just move the oils around or thin them out. If allowed to remain on the surface of unprotected copper alloys, they will etch into the surface and require the use of mild abrasives to remove the stain and restore the finish.

The protected indoor or rural outdoor environments of category C2 will result in alloy behaviors similar to those of the C1 environments. The surfaces of unprotected copper alloys will tarnish over time as oxides develop on the surface. Surfaces should be well drained if they remain outdoors and kept dry when they are indoors. Cleaning fluids, if they are used, should be completely removed from all surfaces and edges. Figure 7.18 shows an example of a surface on which cleaning fluids or polishing compounds remained in the seams or along the edges, causing the brass to corrode. The damage is more esthetic than detrimental to the copper alloy. Still, dezincification of the brass alloy could develop over time. Solder joints should be clean and free of all residual flux. If not, they will slowly develop spots or edges of corrosion as moisture from the air activates them.

In the outdoor environments of C3, which include clean urban environments and those that experience temperature and humidity changes, alloys will darken over time. Water streaks may be

FIGURE 7.18 Cleaning compounds left on the surface.

present and deposits from collected dirt and oils may accumulate in certain areas, creating a minor esthetic concern. Unprotected copper alloys may darken or lighten where these collections occur.

More troubling conditions develop when corroded steel or iron in proximity is allowed to drain on to the copper alloy surface. There most likely is galvanic corrosion contributing to the corrosion, such as shown in Figure 7.19. These corrosion particles will add to the mineral makeup and can be difficult to remove without removing the patina on the copper alloy surface with it.

For all exterior and unconditioned spaces, the dew point can play a role in the development of light oxides on the surfaces. The relative humidity adds minute levels of moisture to the surface and this will react with foreign substances on the surface. Even in arid exposures, the lag in the temperature gradient between the metal surface and the surrounding air as the morning sun heats up the ambient air will result in minute particles of moisture condensing on the cooler metal. This moisture will accelerate the development of an oxide layer on the surface. See the graph in Figure 7.20. This indicates the lag in temperature of the ambient air and the metal.

This tarnish will continue to darken the copper alloy, creating a brownish layer over the entire surface. On polished brasses the tarnish will be streaky and uneven in others handprints will create ideal zones for cuprous oxide film to form (Figure 7.17).

At this point, the tarnish can remain. It is a darkened haze that resembles a statuary finish. Copper roofing or wall surfaces will eventually lose their sheen, while interior brass surfaces will look blotchy and irregular, requiring cleaning and polishing to return them to their original beauty.

FIGURE 7.19 Corroding steel support under a cast copper alloy handrail cap.

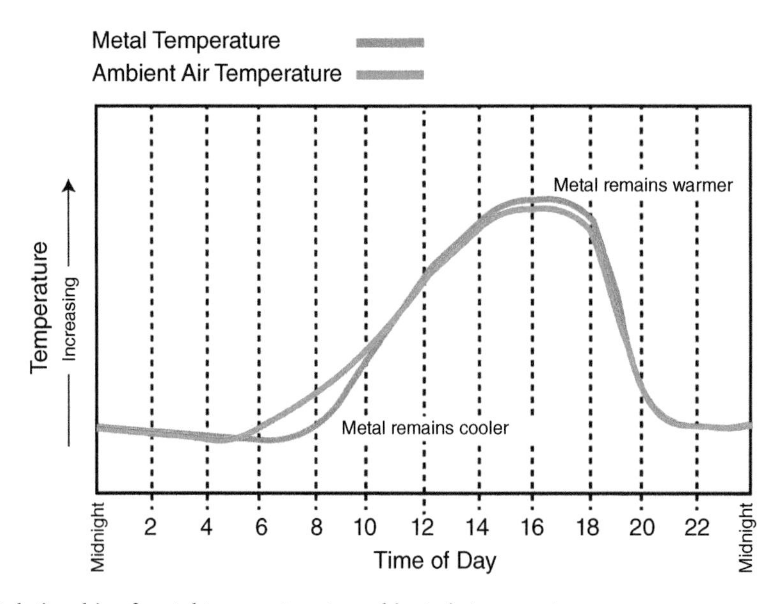

FIGURE 7.20 Relationship of metal temperature to ambient air temperature.

ACIDS AND BASES

Copper will be attacked by nitric acid but not by hydrochloric acid or sulfuric acid. Hydrochloric acid will cause the copper to develop cupric chloride on the surface. Acids used as solder or welding fluxes will develop corrosion cells that will lead to the formation of copper salts on copper alloy surfaces. Figure 7.21 shows a 100-year-old soldered corner on a copper time capsule. The box is made of copper and the corners were cut from Cartridge Brass (alloy C26000). The solder used was a 50-50 tin and lead solder. The flux was most likely created by adding zinc strips to diluted hydrochloric acid (muriatic acid), which formed zinc chloride, a powerful flux used by metal workers for decades.

Some of the flux remained on the surface or leached out from under the brass corner. You can see various corrosion products across the box and around the seams. These are chlorides from the flux residue. The time capsule had been placed inside a stone wall, so the area would have been cool and for the most part dry. Copper corrosion develops quickly but then slows way down as the corrosion products act to protect the metal. The copper itself has a fine layer of reddish cuprous oxide over the surface.

Strong bases such as sodium hydroxide will not attack copper, but ammonium hydroxide will dissolve copper and form a bluish solution. Wood substrates treated with ammonium hydroxide will attack the copper surface. When the wood is moistened, there is outgassing of the wood and corrosion products will appear on the copper surface.

FIGURE 7.21 One-hundred-year-old copper and brass time capsule.

FIGURE 7.22 Copper panels affected by moisture entering a crate.

Figure 7.22 shows 2-millimeter copper panels with a custom red patina. These were shipped in wooden crates. The crates were export grade, so the wood had all been kiln dried or treated for potential insect infestation.[4]

The crates were exposed to moisture for several weeks. The crates had been set outside and rains had soaked the wooden crate. The panels were intended for the outside walls of a museum—if they are destined for the exterior, then why not store them outside, was the misguided thought. The humidity in the crate coupled with the substances that leached from the wood caused rapid surface corrosion. The finish on the copper panels was ruined and had to be removed and re-created.

The difficulty with removing the stains on an oxide's surface, whether iron oxide (rust stains), surface reactions to moisture leaching from wood, or oxidation streaks forming on sculpture, is that removing them will also require the removing of the original patina. The stains are integrated into the oxide compounds that make up the patina (Figure 7.23). These are conditions where the surface patina has been chemically altered, often producing corrosion compounds that have combined with the copper mineral substances on the surface. They are not easily repaired without taking the surface back to the base metal and recreating the patina.

[4]Export-grade lumber is either heat-treated or fumigated to be certain there are no insect infestations in the lumber. There is a grading system and a stamp indicating compliance. They typically do not use ammonium based compounds.

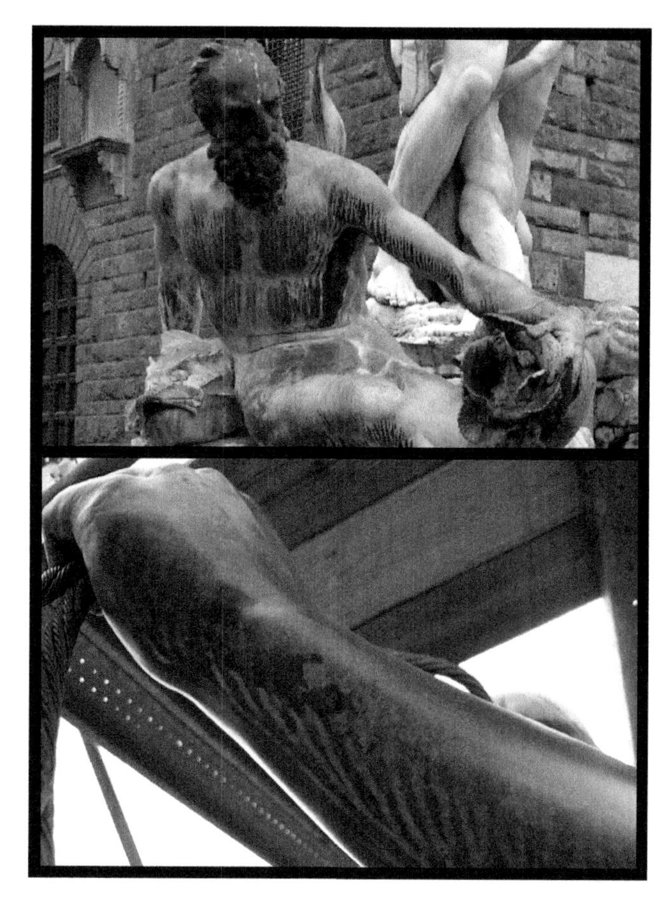

FIGURE 7.23 Streaks on the upper regions of a sculpture.

The C4 and C5 exposure categories correspond to environments in which pollution compounds will form on unprotected copper alloys or where the protective layer gives way to allow the pollutants to attack the copper alloy surface.

There are two main vehicles that will attack copper alloys used in art and architecture: sulfides and chlorides. Less frequently, there can be conditions where nitrates and phosphates from fertilizers attack surfaces and form copper salts.

The chlorides in deicing salts will combine with copper and form copper chloride compounds that have a distinctive chalky green color (Figure 7.24). Usually the stain is streaky and spotty, and its esthetic character can damage the overall appearance of a piece. It can be removed if addressed quickly, but often its removal will brighten the stained area, as some of the oxidation that was on the copper alloy originally will be removed with the stain. Chapter 8 goes into more detail on how to prevent a stain from occurring and how to clean it.

FIGURE 7.24 The chlorides in deicing salts will affect copper surfaces.

COPPER ALLOY SURFACE CATEGORIES

There are three types of surfaces that lie at the interface of the base metal and the environment. They are:

1. Uncoated base metal surfaces
2. Patinated metal surfaces
3. Coated metal/coated patinated metal

There is a distinct difference between these categories and how they will behave in an environment. Alloying can play a role and maintenance will play a role in the expectations for their appearance.

Uncoated Base Metal Surfaces

This category includes the common copper roof and copper wall surfaces. The intention is for the metal to be allowed to interact with the environment. Normally copper roof and copper wall surfaces are allowed to darken and change with time and exposure. Nothing more is added to the metal surface. The metal moves from bright, shiny, newly installed copper to an intermediate phase where interference oxides form on the surface to an eventual dark brown color. This natural progression will occur as the copper absorbs moisture, carbon dioxide, and, if present, chlorides and sulfides. The amount of time that the changes take depends on the relative humidity and the length of time that conditions are wet. The left-hand image in Figure 7.25 depicts a copper alloy storefront wall in Munich. The copper alloy contains zinc and tin. As the surface is exposed to natural weathering cycles, the surface oxide grows to a rich dark reddish brown. Note the light color streak in the image coincides with the joint in the upper sill below the window. This is where water collects, and the joint inadvertently directs it down the face. This is a design detail issue, most likely not intentional.

FIGURE 7.25 Uncoated copper roof and walls.

The problem could have been prevented by relocating the joint and increasing the slope of the top to prevent water from collecting.

The upper right-hand image in Figure 7.25 is the roof of the de Young museum in Golden Gate Park, San Francisco. The photo was taken after about three years of exposure. This massive roof of copper is turning a deep rich brown. If viewed today you would see a tint of green from reactions with the chloride deposits of the salt fog that rolls in from San Francisco Bay. At the lower right is an image of a newly installed copper roof in Minnesota. The bright copper appearance will soon give way to a rich brown tone.

Copper alloys used in interior applications that have just been given a polish or finished to pull out their inherent rich colors are also in this category. In this case, the desire is for the surface not to change and to provide an esthetic experience based on the inherent color of the copper alloy.

The brass alloy floor inserts shown in Figure 7.26 and used in the terrazzo floor of Miami International Airport are an example of an interior copper alloy surface. Cut using waterjet, they keep their bright yellow color through the constant abrasion of foot traffic and the periodic maintenance of floor polishing operations performed by the airport authority.

In most instances in which the rich color is an esthetic design feature, a thin, clear barrier coating is applied to copper alloy surfaces. Without a barrier coating to keep fingerprints and

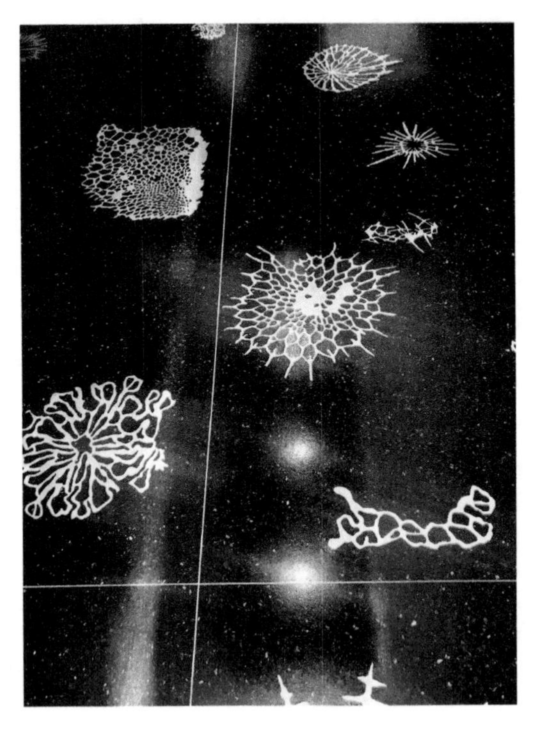

FIGURE 7.26 Copper alloy art inserted in the terrazzo floor of Miami International Airport.

moisture from reaching the surface, tarnish and oxidation will occur and a maintenance regimen will be required.

Patinated Metal Surfaces

In this category an artificial patina of enhanced color will have been produced on the copper alloy surface. Assuming it is sound and that the necessary bond with the underlying metal has been created, the artificial patina will provide the outer layer of protection. The patina is considered artificial because the chemical reaction that creates the color is induced artificially at a plant or in a laboratory.

For large-scale architecture projects, the process of producing an artificial patina is not simple and is prone to failure if not performed by those versed in the art. In addition, on large-scale architecture projects the chemicals used can pose hazards to water supplies and therefore should be performed by reputable companies with established waste- and rinse-water collection processes. In this way little waste will be produced, as the metals and chemical compounds will be collected and recycled or neutralized and disposed of in appropriate ways.

Prepatinated copper roofs and walls with the characteristic green verde surfaces are simulations of what would happen over decades of exposure. These colorful surfaces look and feel very much like a naturally formed patina. Their chemical makeup is similar to a natural patina, being composed of copper sulfate or copper chloride compounds chemically bound to the base metal. There is often an intermediate, or conversion, layer that assists in the chemical reaction and binding of the patina to the base copper. When done properly, these surfaces should continue to slowly respond to the environment and develop a tight bond with the base metal.

These patinas offer excellent corrosion barriers as well, since they represent chemically stable barriers. Oxygen and water slowly diffuse through the coatings, making them tightly bound to the copper base metal (Figure 7.27).

The failure of an accelerated patina usually takes the form either of its darkening to blackish tones or lifting off the surface. Darkening can occur on poorly drained patinated surfaces where the chemistry used to create the patina in the first place is reactivated. Often there is an intermediate oxide under the green-colored copper compounds. This oxide is black and forms a tight barrier to the copper alloy surface; the patina is actually developed over this darkened intermediate layer. When moisture laden with electrolytes from airborne substances is allowed to remain on the surface, it can cause the outer layer to lift, exposing the darkened layer. The top image in Figure 7.28 reveals the dark underlayer below the green patina layer of this wall surface. This is not a failure of the patina, and this layer will facilitate the deepening of the copper sulfate with time and exposure. Note that the color is a deep green, which indicates the patina solution is sulfide based. The bottom image in Figure 7.28, however, shows a situation in which moisture was allowed to collect on a surface. This is a horizontal surface in an urban environment. You can see how the moisture collected and pooled by the nature of the lines. The darkened intermediate layer is visible, as are areas where the copper itself is visible. This indicates the outer patina layer has lifted from the surface.

Green and blue green are not the only durable colors that can be used on copper alloys without additional barrier coatings. Black oxides are extremely durable, while red oxides and the variegated

FIGURE 7.27 (*Left*) Patinated copper art, (*middle*) roofs, and (*right*) walls.

oxide sometimes referred to as Dirty Penny will slowly change and darken over time as cuprous oxide forms. Still, these oxides provide a layer of protection to the copper base metal and slow the effects of atmospheric aging. Note, however, that the presence of chlorides and sulfides will cause a color transformation as these elements along with moisture combine with the copper oxides and develop the more stable copper sulfates and chlorides. The rate of the development of these compounds and the color changes are dependent on the exposure. Seaside and deicing salts will cause changes in one season of exposure.

In sculpture, the cast copper alloys of silicon bronze or brass are frequently coated with an additional barrier of an organic clear film. There are some patinas applied to cast copper alloys that are more durable than others and that on occasion are left to weather without an additional film. There are the copper sulfate green and blue-green patinas described for copper sheet in Chapter 3. As discussed in the same chapter, there are dark brown to dark gray patinas as well; these are produced with ferric nitrate or potassium sulfide, respectively.

In the left-hand image in Figure 7.29, the statue of the young girl has a durable green patina. The black streaks are common and are a sign that cupric oxide is forming on the surface. These

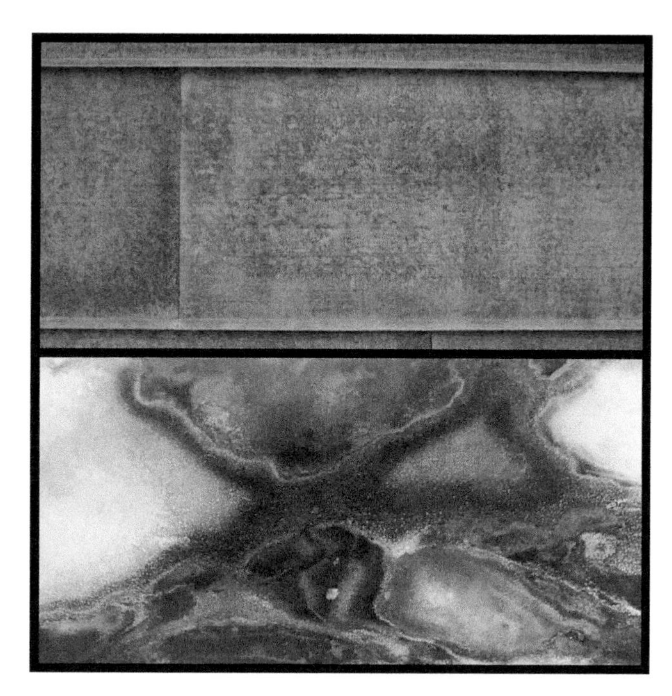

FIGURE 7.28 Prepatinated copper surfaces: (*top*) successful and (*bottom*) unsuccessful.

contrasting streaks do not diminish the appearance of the sculpture. They formed over time as water trickled over the surface. Along with the patina, these dark tones give the sculpture a timeless beauty and are fully expected to occur.

In the right-hand image, the durable gray patina shown was produced by the foundry with potassium sulfide and applied to cast silicon bronze. This patina will not be as durable as the green patina produced with copper sulfide, and over time it will fade slightly, as you can see from the lightening around the collar. If washed frequently to remove bird waste and airborne particles this sculpture and its patina would perform well unprotected, but a coating of wax or protective lacquer would keep the sculpture looking as it was newly installed.

Coated Metal and Coated Patinated Metal

On cast sculpture, most surfaces are coated with an additional barrier. On architectural surfaces, the more delicate oxides, such as the statuary finishes, are further coated to protect them, even in interior applications. There are, additionally, those occasions where the natural, unoxidized color is desirable and the metal is coated to maintain the color. These coatings and their expected performance are described in more detail in Chapter 8.

There are conditions, however, where corrosion of the copper alloys is influenced by the coatings used and the way that they are applied. Because of the excellent corrosion resistance of copper and

FIGURE 7.29 Unprotected patinas on bronze.

copper alloys, the service utility for the useful life span of the casting or copper alloy surface will not be affected. The esthetic appearance, however, will.

After years of exposure, a beautifully patinated bronze sculpture can appear shabby and worn. A polished brass surface can develop thick oxides and darkening around the edges if the protective barrier gives way.

The most common problem occurs when a clear protective coating is applied over a copper alloy surface that still has trace moisture, oils, or fingerprints on the surface. The clear coating may not reveal moisture or oil under the surface at first, but as the coating cures reactions with these films create initially small areas of dark tarnish under the film and on the surface of the coated copper alloy.

These dark spots grow and can form large dark regions, as shown in Figure 7.30. In this instance, a sulfide finish was applied to darken the engraved regions where the lettering was cut into the C28000 brass. To remove excess darkening, the surface was polished again to remove the darkening and leave it only in the recessed areas. This creates the contrast and allows the lettering and design to be expressed to the viewer. During this process the surface went through several rinse cycles and

FIGURE 7.30 Under-film oxidation on brass.

dry cycles. If the surface is cold it may retain traces of moisture and these will slowly react with the copper alloy, causing dark tones to appear over time.

Fingerprints are notorious. Invisible initially, they show over time as the mild fatty acids react with the copper alloy. The fingerprint or handprint can appear as a ghost remnant under the film. Correcting this condition requires removing the clear coating, recleaning the surface, and reapplying the clear coating.

> Copper and copper alloys are some of the most corrosion resistant metals used in art and architecture. For centuries they have proven time and time again their resilience in challenging atmospheric exposures, whether polluted urban environments or destructive humid seaside conditions. The surface of the metal will change. This change is what gives copper alloys their toughness—and in many cases, their inherent beauty.

CHAPTER 8

Maintaining the Copper Alloy Surface

The vessel, though her mast be firm, beneath her copper bears a worm.

Henry David Thoreau

INTRODUCTION

The alloys of copper are some of the most corrosion resistant metals in common use today. They achieve this by rapidly producing a layer of inert corrosion products on the surface. In the context of art and architecture, however, this tendency for rapid change in surface appearance and luster can be daunting. Unprotected copper alloys and those whose coatings have been breached or stained will change in appearance, often to a form that is less aesthetically pleasing.

The surfaces of copper and copper alloys undergo a change in their morphology, meaning that they will change in appearance as oxidation proceeds until a stability is achieved. For copper alloys with significant zinc content, this change in morphology can lead to physical degradation of the metal itself as zinc is selectively removed in the corrosion phenomenon known as dezincification. Bronze sculptures, particularly when exposed to chlorides, can develop what has been referred to as bronze disease, in which the surface is selectively attacked. The surface will continue to corrode if the chloride exposure is not contained.

Much of what occurs, if addressed early, can be removed, stopped, or restored. Keeping surfaces clean requires either protecting the surface with a substance such as a wax or lacquer or performing regular surface cleaning to remove foreign substances that have found their way to the surface.

PROTECTING THE NEW COPPER ALLOY SURFACE

When copper alloy surfaces are expected to be displayed with their natural metal luster, they will either require a coating of some form or another or regular cleaning and polishing. The coating will usually be either an organic coating (such as a clear lacquer or wax) or an inorganic coating (such as a chromate or silane coating). Inorganic coatings are not as commonly used in art and architecture, and are usually considered when thermal conditions could prematurely damage an organic coating. Organic coatings have a shorter life than inorganic coatings but they usually can be removed and restored, while inorganic coatings can't be restored but have a longer expected life.

One of the biggest challenges with organic clear coatings used to protect copper alloys is the potential development of tarnish under the film. This tarnish, which will develop later in the life of the coated copper alloy, is from peroxide anions that accumulate as the organic coating begins its first stages of degradation. The anions develop early in the life of the coating from trapped moisture below the film or residual solvents decaying under ultraviolet exposure.

As with other metals, one can evaluate the condition of the copper alloy surface from the point of view of these three categories:

1. Physical cleanliness
2. Chemical cleanliness
3. Mechanical cleanliness

ACHIEVING PHYSICAL CLEANLINESS

To be physically clean, the copper alloy surface must be free of tarnish and light oxidation, fingerprints, soils, and other substances that have yet to engage with the base metal itself. If an artificial patina is to be generated on the surface of the copper alloy, the surface must first be physically clean. A physically clean surface is the condition nearest to a new metal surface: free of and unfettered by coatings or oxides.

Aspects of Physical Cleanliness for the Copper Alloy Surface

- Oxide free
- No tarnish
- No fingerprints
- No soils
- No oils
- No adhesives

Tarnish

All of the copper alloys will develop a light oxide on the surface when exposed to the air. This oxide will continue to grow. The oxide is a form of cuprous oxide (Cu_2O).

Figure 8.1 shows the tarnished surface of a door made from copper plate. The surface is exposed to the outside environment and has developed a light oxide. The colors seen are induced from the light interference effects of the thin film.

This tarnish forms on the surface of clean, unprotected copper alloy surfaces with the exception of aluminum–bronze alloys. The aluminum-bronze alloys form an aluminum oxide on their surfaces and are resistant to tarnishing. Tarnish is the natural first step toward mineralization of the copper surface as oxygen and hydroxides combine with the copper surface. They are weakly bonded at first so they can be removed more easily.

The tarnish can be removed using mild oxide removers designed for copper alloys. There are numerous tarnish removers for copper, copper alloys, and silver that work quickly and effectively. It is critical, however, to remove all remnants of the cleaning solution from the surface and from the recesses and seams of a fabricated assembly. If the cleaning solutions are allowed to remain in the seams, they can etch the surface and develop corrosion conditions favorable to corrosion. Commercial tarnish cleaners are composed of acids, bases, or neutral pH cleaners. Some have mild abrasives and assist in polishing the surface. It is critical to remove all remnants of the cleaners, otherwise they can continue to corrode the copper alloy. Alkaline-based cleaners will remove tarnish well, but they also can corrode the copper if they remain on the surface. The corrosion may be superficial, but subsequent cleaning to remove the corrosion products will be more difficult.

FIGURE 8.1 Tarnish on copper surface of a door made from thick copper plates.

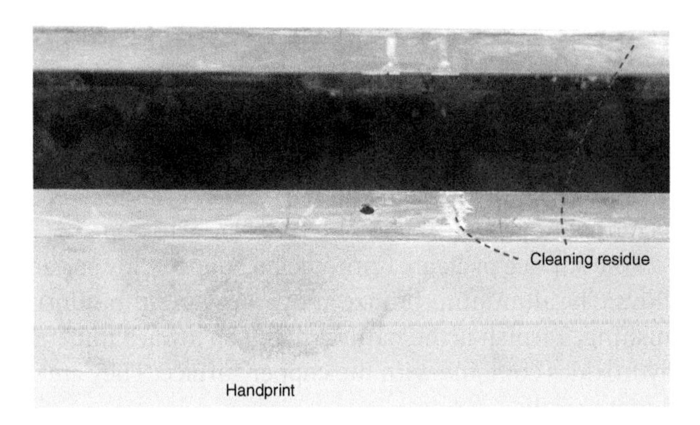

FIGURE 8.2 Cleaning solution and handprint left on a copper alloy form.

If the pieces are cleaned and placed in storage, it is even more critical to ensure that cleaning fluids and pastes have been thoroughly removed. These will continue to react with the copper surface and cause corrosion. Worse yet, they can create conditions of dezincification in the high-zinc brass alloys.

Figure 8.2 shows remnants of cleaning fluids that have seeped onto the reverse side of a polished brass form. A visible handprint has oxidized the surface of the copper alloy.

Fingerprints

Clean, uncoated copper alloys fingerprint readily. The oils left behind by hand- and fingerprints are composed of organic oils and fats called lipids, which are composed of amino acids and water. They will lightly etch the surface and create a contrasting appearance. Often they will not be readily apparent until the relative humidity rises. They will not disappear on their own (Figure 8.3).

It is possible to remove this light oxidation from the surface of copper and copper alloys where the design called for a pristine, natural appearance by using commercial cleaners designed for copper. Some buffing and polishing may be beneficial to improve the gloss and shine of the original appearance. Coat them with wax or clear lacquer to sustain the appearance is recommended; otherwise, plan on ongoing maintenance to remove tarnish that develops.

Under-Film Corrosion

Tarnish and handprints can also appear under the clear coatings applied to protect the copper alloy surface. This is due to moisture or other substances being on the metal prior to the application of the clear lacquer. The solvents in the lacquer can also undergo a transformation and cause an oxidation reaction. It can be a significant and costly problem. Initially the corrosion may not be visible, but it will grow with time and exposure.

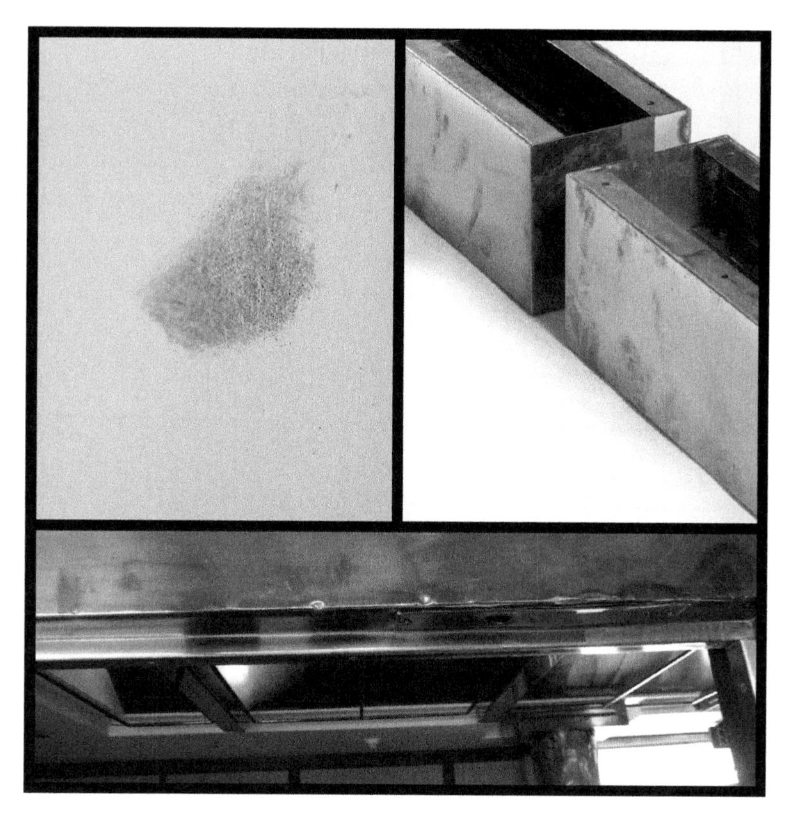

FIGURE 8.3 Fingerprints on copper alloy surfaces.

Figure 8.4 shows brass plaques that were highlighted with a light statuary finish to darken the etched portions. As time passed and the temperature changed, the panels darkened under the film. Deicing salts left on the lacquer could also have been a contributing factor. Any porosity in the film would have allowed the small chloride atom to enter and react with copper alloy surface.

This can occur even with lacquers containing antioxidation additions such as sodium benzotriazole. This is because the oil, moisture, or handprint was between the copper alloy surface and the antioxidation compounds in the lacquer. The darkening begins as spots under the lacquer as shown in the right-hand image in Figure 8.4.The only means of correcting this is to strip the protective coating from the surface and remove the oxidation. Removing the coating on large panels and objects can be a daunting challenge when lacquers or varnishes are used. Removing these will require a solvent capable of dissolving the lacquer. It may require that the panels be removed and soaked in a bath of the solvent. Disposal, health, and flammability issues are primary concerns to address before attempting this.

In order to combat the return of tarnish or under-film degradation, or to prevent its development in the first place, thoroughly clean the surface once the lacquer is removed prior to coating

FIGURE 8.4 Panels at the 9/11 Memorial darkening under the film.

or recoating the surface. Heat it slightly to drive moisture out of the copper alloy surface pores. Do not overheat it or it will oxidize when it cools. Heating will not damage the statuary finish or patina. Wipe the surface down with an inhibitor solution containing sodium benzotriazole. A solution with 1–2% sodium benzotriazole and distilled water will work, or a 2% solution of benzotriazole and alcohol will keep moisture out and not affect the lacquer application. Use a lacquer coating that has ultraviolet absorbers and antioxidation enhancers in the lacquer. Incralac, an acrylic lacquer developed from research sponsored by INCRA, is an excellent lacquer that is used on copper alloys for this purpose.

Incralac has ultraviolet absorbers and benzotriazole included with an acrylic ester resin dissolved in a solvent.

The ability of benzotriazole to prevent tarnish from developing on copper alloys was discussed in Chapter 4, and an explanation of what is occurring can be found in the same chapter. Figure 8.5 shows the results of a test in which a C11000 copper tube was first thoroughly cleaned, after which one half of the length was coated with a 2% sodium benzotriazole solution, allowed to dry, and then dipped in a strong solution of potassium sulfide. The unprotected portion of the tube immediately darkened while the half that was coated remained unaffected.

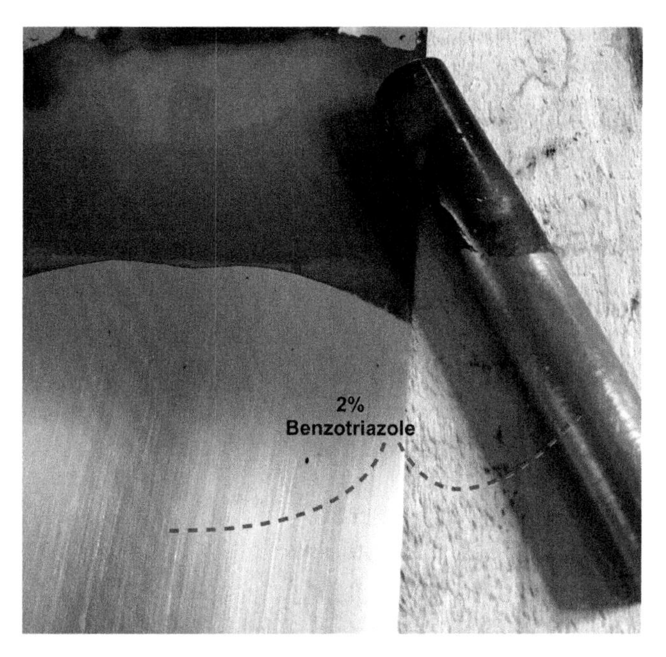

FIGURE 8.5 Image of test results of ability of benzotriazole to prevent tarnish.

When a copper alloy part is coated with the sodium benzotriazole, a thin layer is deposited on the clean surface. The sodium benzotriazole coating is imperceptible to the eye once the surface is dry. It has been used as a protective oxide inhibitor in museum work and it is a common oxide inhibitor used by the industry when shipping copper alloys to be further processed in the fabrication chain.

Eventually the benzotriazole begins to wear off and tarnish will occur. Thus, using it alone will require repeated treatments; otherwise, use it with a wax or Incralac coating.

Other Substances on the Surface

Bird Waste

On exterior copper surfaces, in particular sculpture, there are several common substances that inflict damage on the surface of the metal. It is difficult to envision a sculpture without bird waste on the upper regions. Bird waste is a watery paste composed of large amounts of nitrogen and phosphorus. Birds do not produce urine, so their waste is a pasty watery substance composed of uric acid. Uric acid is for the most part insoluble and will adhere to the surface. It can become a baked-on crust as the sun heats the surface and decompose into components that will react with the copper alloy surface and create streaks of greening corrosion products. It can also adhere to old wax and create stains on the sculpture (Figure 8.6).

FIGURE 8.6 Bird waste on copper wall panels and a bronze sculpture.

To prevent this, bird waste needs to be cleaned from the surface regularly. To break the adhesion, pressure washing the surface or physically scrubbing the surface is advised. Detergents can assist in lifting the waste off the surface, but the putty-like substances need physical displacement.

If allowed to remain they will discolor the surface and damage the underlying patina as they decompose into ammonia and phosphorus-based substances that will combine with the copper. Figure 8.7 shows a bronze and stainless steel outdoor sculpture designed by the artist James Surls and titled *Six and Seven Flowers*. The horizontal "arms" are perfect perches for the birds in the area. You can see where the bird waste is beginning to interact with the copper alloy to create a slight green tint.

If the bronze petals were protected with a clear lacquer and wax, it would be easy to remove the bird waste with an occasionally pressure rinse. Allowing the bird waste to interact with the copper alloy will create corrosion products that will demand more difficult cleaning. The corrosion particles will need to be dissolved, unfortunately, along with the original patina. This will require the restoration of the patina after cleaning.

FIGURE 8.7 Bird waste beginning to interact with the copper alloy used for
Source: *Six and Seven Flowers* by the artist James Surls.

Other substances that will affect the physical cleanliness of a copper surface are adhesives and glues. General dirt and grime would fall into this category, as well as old wax coatings. Even printing by the Mill on the copper surface before patination will affect the end appearance, and thus the need for physical cleaning, before moving to the steps of patination. Adhesives, glues, inks, and old wax can be removed with commercial solvents such as mineral spirits or naphtha.

Removing potentially damaging substances from the surface, whether on bare metal, patinated metal, or coated metal, is advised for good long-term performance. Figure 8.8 shows conditions that need to be addressed to bring the copper alloy back to a level of physical cleanliness. The condition of printing, as in the image at the lower right in Figure 8.8, will require removing the patina or starting over with physically clean copper.

Deicing Salts

The use of deicing salts in northern climates is very common. Unfortunately, these salts find their way onto the surface of metals in proximity of walkways and doorways. Deicing salts are composed

FIGURE 8.8 (*Left*) Bird waste, (*top right*) old wax, and (*bottom right*) Mill printing under the patina.

of chloride salts such as magnesium, calcium, and sodium chloride. When the salts are applied to icy surfaces, they melt the ice by lowering the freezing point of water, a chemical behavior referred to as "freezing point depression." The salts dissolve in the water and become ions of chloride and sodium or chloride and calcium. As more particles are added, it becomes more difficult for the water to freeze. It is the chloride ion in the water that causes stains on copper if it is splashed onto the surface and allowed to remain there. In the cold of winter, when the salts are applied, the copper and chloride are slow to react. The telltale sign of deicing salts on a surface is a white stain, as shown in Figure 8.9. The stain adheres to the metal surface and can be difficult to remove. To remove it, a deionized water rinse is recommended. A light pressure wash will assist in the removal as well.

As the higher temperatures of spring warm the surface and dew condensates on the cooler metal in the morning hours, the chloride and the copper will react and form copper chloride. Figure 8.10 shows three surfaces where the chloride and the copper have started to react and form copper chloride on the surface. These stains are mineral forms of copper now and cannot be removed by pressure washing, deionized water, or solvents.

FIGURE 8.9 Deicing salt residue on surfaces.

It is always the lower sections of copper alloy surfaces that show the corrosion product copper chloride ($CuCl_2$), a turquoise bluish green hydrated mineral. The stain is superficial and does little harm to the copper other than making an esthetic alteration to the color. The difficulty is in removal, because removing the stain will remove all oxide from the copper surface. You cannot remove the stain without removing the oxide on the base metal. It is a patina, a mineralized surface that will require either physical removal down to the base metal, dissolving in acid, or basic solution treatments. Once clean, the surface will need to be reoxidize to restore the original surface color.

Waxing and lacquers can protect the surface of copper alloys for a while, or at least long enough for maintenance processes to remove all surface chlorides as indicated by the salt residue. Deionized water will effectively remove ions of chloride and other ionic forms that are on the surface. Thoroughly rinsing the surface in early spring before the chloride and copper have time to react is highly recommended. If the surfaces are coated with wax or clear lacquers, inspect the coatings and do any necessary repairs. You can usually find areas where the coating has been breached indicated by spots of darkening.

FIGURE 8.10 Deicing salt damage to copper surfaces.

Old Wax

On sculpture more so than architectural surfaces, old wax can make a surface look drab and washed out. Figure 8.11 shows a surface of old wax that has yellowed in the left-hand image and the same surface brightened, protected, and enhanced by new wax on the right.

The old wax is easy to remove, and as it comes off many potentially damaging substances will be removed with it. Removing old wax offers a chance to closely inspect the surface as well. Remove old wax by wiping the surface with clean rags soaked in mineral spirits. Mineral spirits are mild aliphatic solvents, but working with them still requires protection of skin and eyes as well as surrounding areas. Avoid breathing in the fumes as much as possible, and use them only in a ventilated area away from open flames.

There are different levels of quality of mineral spirits, which are also called white spirits or naphtha. It is important to use the better grades. Removing the wax will be easier and less messy. Museum grade mineral spirits work exceptionally well.

Once the old wax is removed, it is good practice to clean the surface with mild detergent and water. Allow the surface to dry thoroughly before applying the new wax. The use of a torch

FIGURE 8.11 (*Left*) Old wax and (*right*) renewed wax on a bronze sculpture by the artist
Source: Kwan Wu.

helps to ensure that water is out of the surface pores of the metal, and gently heating the surface promotes the flowing of the new wax into tight regions. Use a high-grade wax that has ultraviolet inhibitors and dry it to a hard, durable coating. It is best to apply the wax in layers to insure good coverage. Later in this chapter there is an in-depth discussion of the waxes used on copper alloys.

Old Lacquers

Removing old lacquer coatings on copper surfaces can be challenging. The first problem is determining what exactly you are removing. Lacquer is an eclectic term covering a wide range of possible coatings. Acrylic-based lacquers (such as Incralac) have been in common use on sculpture and most copper alloys since the 1970s. The makeup of Incralac and its use is described in more detail later in this chapter.

Other lacquers have also been used with varying levels of success, linseed oil, urethanes, and in more recent times, sol-gel applications, have been used. Sol-gel applications are showing promise as coatings for metals. Sol-gel is a process of applying thin films of discrete particles of metal oxides, either silicon or titanium oxide, over the surface.

Prior to the 1970s, sculpture was often coated with a cellulose-based compound. One such compound was cellulose acetate butylate. It was a coating used on some bronzes, particularly those exposed to a lot of moisture, such as artistic fountains. An example of this coating is shown in Figure 8.12. This sculpture, called the *Muse of Missouri*, was constructed in the early 1960s. It was designed by the artist Wheeler Williams and cast at the Modern Art Foundry of New York. It is 4.5 m tall and stands on top of a pedestal fountain 3.9 m tall. It has been cleaned and treated several times over the last 50 years, but the original coating protecting the patina on the pedestal had never been removed.

FIGURE 8.12 *Muse of Missouri* by Wheeler Williams.

The image on the upper right shows the whitish residue that had developed over the years. It was determined by several tests to be decomposed cellulose nitrate compound that had formed a thick yellowish crust from the decaying of the cellulose acetate butylate. Figure 8.13 shows a microscopic image of the surface on the left before removal of the crust and after removal on the right. Hard-water deposits had also formed on the surface. The compound was very durable and could not be dissolved by treatments with strong solvents such as acetone or xylene or with acids such as phosphoric, citric, or gluconic acids. As the compound decayed, it had formed a plastic-like substance on the surface.

Ethyl acetate was eventually used to dissolve the cellulose substance. Ethyl acetate soaked rags held onto the surface were found to soften the surface and allow for the substance to be removed. Refer to the lower right image of Figure 8.12. Below the crust the patina was in good shape and could be protected with Incralac and wax. For fountains, three layers of Incralac followed by a wax coating have been found to provide very good protection to the surface.

Stains from substances in the fountain's water, in particular rust, had also deposited on the surface and created appearance issues. The stains came off when the underlaying coating was removed.

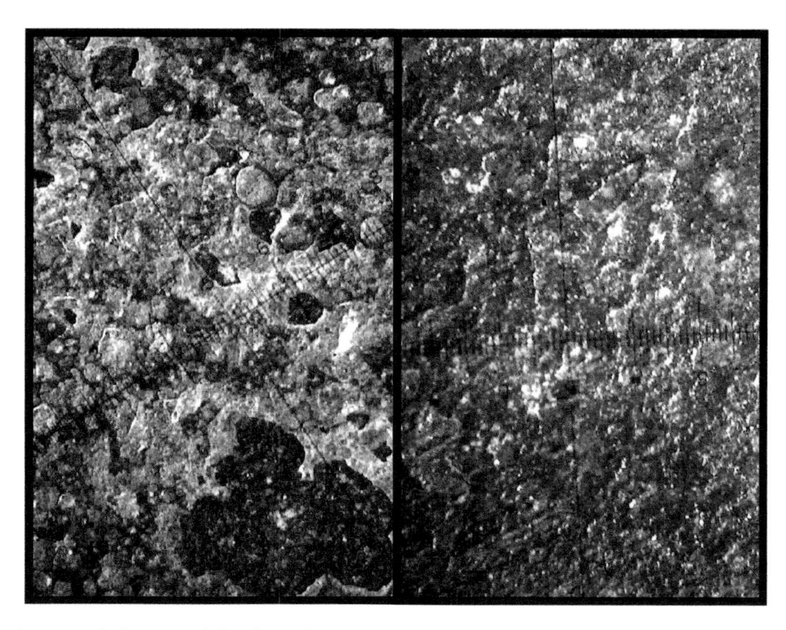

FIGURE 8.13 Microscopic image of the decaying coating.

Subsequent removal of the wax will remove rust stains with it, allowing the sculpture to look and perform as it did when first installed.

Figure 8.14 shows rust deposits on the copper alloy surface around the feet of the sculpture and the outlet where the water is emitted. Most likely these deposits are from corroding internal pipe fittings on the steel pipe used to deliver water for the fountain.

The approach to cleaning surfaces should be to start simple and work up to more complex cleaning approaches. As a first step, isopropyl alcohol can be used to wipe the surface to remove light soils, light grease, and fingerprints that have not yet interacted with the copper surface to develop light oxidation. This can be followed by cleaning with a mild, neutral detergent and a clean-water sponge or wipe (Figure 8.15). Add pressure, if necessary, to dislodge adhered waste such as bird droppings or road grime. Note that high pressure can potentially damage a wax coating, so use this cautiously on sculpture work.

The point is to remove as much of the general dirt and grime, bird waste, and other potential reactive substances from the surface as possible. As discussed earlier, if deicing salts are left on the surface, they will eventually reach the copper metal and react. Bird droppings have an organic acidic nature that will also eventually react with the copper if allowed to remain.

A final rinse with deionized water is also good practice. The deionized water will remove salts from surfaces. Chlorides, sulfides, and nitrates will come off with a deionized rinse, and as the surface dries spotting will be reduced.

To remove old wax, use mineral spirits, which is a common name for naphtha, or better still odorless mineral spirits, which is safer to use due to the removal of aromatics. Mineral spirits will

FIGURE 8.14 Rust stains on the *Muse* sculpture from deposits in the water.

FIGURE 8.15 Cleaning copper alloy surfaces with mild soap and water.

TABLE 8.1 Levels in achieving physical cleanliness.

Level	Procedure	Benefit
1	Wipe down with isopropyl alcohol	Removes light soils and fingerprints
2	Clean surface with mild detergent	Removes soils and dirt
3	Wipe with commercial tarnish remover (if needed)	Removes light oxide and fingerprint oxides from the surface
4	Pressure wash (if needed)	Removes bird droppings, dirt deposits
5	Mineral spirits (if needed)	Removes old wax, adhesives, glues, inks
6	Degreasers (if needed)	Removes heavy oil and grease deposits
7	Steam (if needed)	Loosen tough soil deposits, bird droppings
8	Deionized water rinse	Removes salts and chemical deposits (use after any lower-level operations as a final step)

dissolve the wax without harming any underlying lacquer or patina. The wax can be removed by wiping the surface with a clean rag saturated with the mineral spirits. After this, new wax can then be reapplied to the surface.

Mineral spirits will also remove most gum, adhesives, and other sticky residues that might find their way to the surface. When removing wax with mineral spirits, it will take with it most substances on the wax. If dirt and grease deposits are heavy, the addition of a steam cleaner may be needed. The added heat will loosen the particles and allow the mineral-spirit wipe to pull these from the surface.

Commercial degreasers can be added if the grease or oil deposits are heavy or are deep into the pores of the metal. These degreasers are often alkaline and must be thoroughly rinsed from the surface, otherwise they will create regions where corrosion reactions can occur. These would show as telltale green deposits on the surface after several weeks of exposure.

Table 8.1 shows cleaning levels for removing substances or addressing various surface issues and their order in the cleaning process.

ACHIEVING CHEMICAL CLEANLINESS

Achieving chemical cleanliness involves removing heavy oxides and stains that have chemically bonded to the copper alloy or copper alloy surface. This includes Mill scale, heat tinting at welds, oxides from cleaning solutions, and solder fluxes that have been left on the surface and have chemically bonded with the metal over time. It also includes free-iron deposits that have been transferred from other surfaces and integrated into the copper oxide in the form of rust stains, as well as deicing salt stains and fertilizer salt stains, which will develop if these substances are allowed to remain on the surface (Figure 8.16).

FIGURE 8.16 (*Right*) Cleaning fluid left on a surface, (*top left*) iron rust stains, (*bottom left*) heavy oxidation.

These substances become a part of the oxide layer on the copper surface. They are an aesthetic issue for the most part because of their contrasting appearance but can become a corrosion concern if conditions of dezincification arise, as is the case in the right-hand image shown in Figure 8.16, where cleaning fluid has been left on a surface. Dezincification can occur on brass alloys with 15% or more zinc. If allowed to remain on the metal surface, all of these substances can lead to light etching of the surface, dezincification corrosion, or pitting.

FIGURE 8.17 Dark streaks of cupric oxide visible on old copper surfaces.

Removal of the contrasting corrosion products will also remove the oxide back to the bare copper alloy. Their removal will require mechanical abrading of the surface or chemical dissolution of the oxide that has formed.

As discussed previously, the copper in copper alloys will oxidize in one of two states. One state is often referred to as copper (I) oxide and has the chemical formula Cu_2O. This compound is known as cuprous oxide and is generally reddish to reddish brown in color. The other form of copper oxide is copper (II) oxide, also known as cupric oxide, and has the chemical formula CuO. This compound is black in color. You can see cupric oxide forming as black streaks on old copper exposed for years (Figure 8.17).

Dissolving the Oxide

Cuprous oxide can often develop on exterior surfaces and can lend a reddish color tone to brass items placed in an interior exposure as well. Citric acid–based cleaners work well on copper alloy to dissolve cupric oxide, but it will not dissolve cuprous oxide. The restoration of several large, intricately designed cast copper alloy doors provides an example of how these oxides may be removed (Figure 8.18). These magnificent doors were cast in the 1920s from alloy C83600. Sometimes referred to as European Bronze, 85-5-5-5, or leaded red brass, the alloy contains nominally 85% copper, 5% zinc, 5% tin, and 5% lead. European Bronze is a cast alloy that was in extensive use in Europe and the United States in the late nineteenth century and early twentieth century.

These doors have been subject to an urban environment exposure and over the years have been exposed to deicing salts. They have also been sand blasted and possibly waxed a few times in their history. The oxides and surface corrosion created an esthetic issue with the owner. The corrosion on the cast surface created areas of golden yellow oxide and a ruddy reddish brown. On inspection there were no signs of dezincification, most likely due to the presence of lead and tin and the low zinc content.

To remove the oxides that had developed on the doors, they were first washed down with mild detergent and rinsed in clean water (Figure 8.19). Then they were dipped into a bath of concentrated

FIGURE 8.18 Cast copper alloy doors after decades of exposure.

FIGURE 8.19 Initial removal of oxides.

lemon juice. Concentrated lemon juice is a source of citric acid. The bath had a pH of 2–2.5. The cast doors were allowed to sit in the bath for several hours.

The lemon juice worked more effectively than phosphoric acid or even a citric acid solution of a similar pH. There is another active component in the organic lemon juice that acted as a surfactant and helped to raise the oxide from the surface. Once the oxide was loosened, the doors were pressure washed with clean water to remove the residue. Figure 8.20 shows the cast copper alloy surface after it was removed from the bath.

Some of the surfaces retained the red cuprous oxide, since lemon juice does not dissolve this from the surface. The color of this copper alloy is golden yellow. To restore the entire surface to

FIGURE 8.20 Much of the tarnish, old oxide, and patina was removed from the surface after a lemon-juice bath.

this color would have involved more potent treatments with phosphoric acid or more powerful acid treatments using sulfuric and nitric acid. Mechanical blasting would also have removed the oxide, but this would have had an undesirable effect on the copper alloy surface. For the application of the final statuary finish such treatments were not needed. The statuary finish would develop over the cuprous oxide and the natural base metal surface without any distinguishing mottling.

Once the oxides were removed, the pieces were wiped down with phosphoric acid and alcohol. Figure 8.21 shows the initial development of a dark oxide treatment on the surface. The image on the right shows the oxide with a whitish film on part of the surface. The image on the left shows the initial application of the finish; some highlighting has been performed, as indicated by the bright edges on the cast features. Highlighting the surface brings out a contrasting appearance and gives an "old bronze" look and feel. The oxide used was a copper sulfide treatment that has a dilute phosphoric acid to act as an etchant and an electrolyte. The color goes dark and sometimes chalky at first. Scotch-Brite pads were used to highlight the surface and remove some of the dark chalkiness.

Once the color was obtained and the highlights applied, the doors were coated with three layers of Incralac. The first coating of Incralac was a mixture of three parts xylene solvent and one part Incralac. This was allowed to dry before a second layer of Incralac, this time two parts xylene to one part Incralac, was applied. This coating was also allowed to fully dry before a third layer, also two parts to one, was applied.

Once the doors were hung (Figure 8.22), they were coated with two coats of wax. The wax used was Trewax® paste wax, which is a blend of carnauba and microcrystalline wax. When applying this wax, allow it to dry between coats and rub it into the surface to even it out. Light heat can be added to help the wax to flow into tight regions around the artwork.

This is one method to chemically clean a surface and then return it to the original patina appearance. There are other methods that involve dissolving oxides from the surface using selective electropolishing techniques. The left-hand image in Figure 8.23 shows the selective electropolishing of a C26000 alloy surface. The oxidation of surface materials and dezincification deposits were removed by passing a selective electropolishing device over the copper alloy surface.

FIGURE 8.21 Developing the statuary finish.

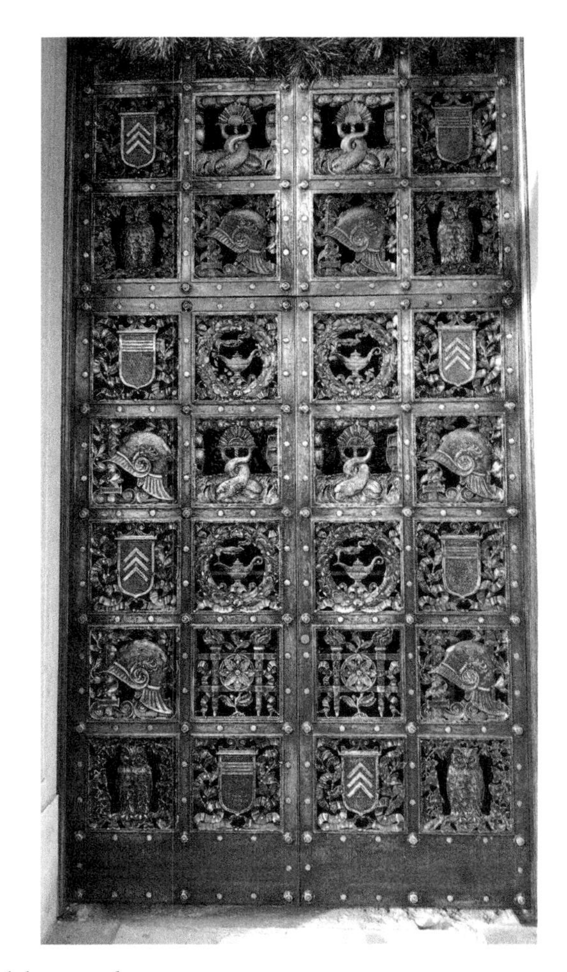

FIGURE 8.22 The restored doors are hung.

FIGURE 8.23 (*Left*) Removing oxides from C26000 alloy and (*right*) repolishing the surface.

Once the oxidation was removed, the surface had a slight etch and needed to be repolished and buffed to bring back the original mirror finish. This is done using rouge coupled with a cotton buffing mop. There is an art to producing the correct polish. It is a messy operation, but buffing and polishing most of the copper alloys are easier than for other metals. The copper alloy surface will take well to buffing and polishing operations.

Electropolishing

Electropolishing and selective electropolishing of copper and copper alloys is an excellent method of removing light oxide deposits and heat tint from welding or brazing.

It is also a good method of removing tarnish and even stains caused by deicing salts.

Electropolishing processes are more commonly used in cleaning and preparing the surface of stainless steels. Copper alloys were some of the first materials to be electropolished in the early decades of the twentieth century. Electropolishing can be performed selectively, as indicated in Figure 8.23, or by immersing the surface in a tank containing an electrolyte, usually phosphoric acid. The copper is made the anode and immersed in a temperature-controlled bath of electrolyte as a current is applied. A cathode is immersed into the solution to complete the circuit. As current is applied from a DC source, the surface protrusions on the copper dissolve by means of oxidation while a reduction process occurs at the cathode and hydrogen is released. The surface oxides are dissolved and high points on the copper alloy are smoothed slightly.

In selective electropolishing, or electrocleaning as it sometimes is called, a wand composed of graphite and wrapped in cotton is the cathode. The cotton is soaked in phosphoric acid and the copper object is connected to the positive terminal of a DC power source. A current is applied and as the wand is passed over the surface, the area under the wand is dissolved. The process of electropolishing copper is the opposite of electroplating : in electroplating the poles are reversed, making the copper electrically positive, immersed in a metal ion solution where the metal comes out of solution and forms onto the copper surface.

There are a number of other methods that can be used to bring the surface of copper alloys back to chemical cleanliness. Mechanical polishing and buffing (Figure 8.24) removes the surface and all oxides with it.

Abrasive Blasting

Other techniques for producing chemical cleanliness involve abrasive blasting of the surface with glass beads or fine sand, dry ice blasting, and blasting with walnut shells. The more aggressive of these cleaning methods, sand blasting, can shape thin sheet material and will create a coarse texture that may detract from the original appearance. This method creates airborne particles that may require special containment and collection.

For thin metals, such as copper roofing or wall sheathing, removing the patina and oxide using blasting techniques is the preferred method. It requires care to not damage the thin metal surfaces. Blasting with walnut shells and dry ice both work well, but dry ice blasting is not as messy.

FIGURE 8.24 Mechanical polishing of a copper alloy surface.

The difficulty of dry ice blasting lies in the fact that the dry ice hopper needs to be near the nozzle of the gun delivering the pellets to the surface. On roofs this may not be possible.

Crushed walnut shells will work and not damage thin metal, but it can place a lot of dust and debris into the air. Collection and disposal of the walnut shells along with the dust particles created is an additional challenge. Both blasting metals are very load. Ear protection is mandatory.

Pickling

Pickling baths composed of sulfuric acid and nitric acid can remove thick oxides from the surface of copper alloys in controlled work areas. Working with these solutions is dangerous and requires a clear safety protocol, an understanding of working with acids, and a proper means of disposal.

You could say that dissolving the oxide on a surface with either lemon juice concentrate, citric acid, or phosphoric acid constitutes a milder form of pickling. These substances are all safer to use than strong acid pickling baths and work well on large surfaces. They do not work as rapidly as the sulfuric and nitric acid mixes, but they are much safer to use and easier to dispose of correctly. On large surfaces, acid baths are impractical.

Laser Ablation

Laser ablation techniques are another way to remove oxides from the surface of copper alloys. These techniques use a vacuum system that draws the fumes and released surface oxides into a filter trap. Laser ablation is a clean method of removing the oxidation and contaminants on copper alloy surfaces. Figure 8.25 shows the removal of patina from a perforated copper surface by means of a 1064-watt laser ablation system. The laser ablation process can be performed in situ and uses no solvents or abrasives. The surface of the copper alloy has a light peening when viewed under a microscope.

FIGURE 8.25 Laser ablation of copper surface.

TABLE 8.2 Chemically clean surface levels for copper alloys.

Levels	Procedure	Benefit
1	Immersion in acid bath	Remove oxides from surface
2	Selective electropolish	Weld tint, oxides, corrosion products
3	Immersion in stronger acids, blasting with dry ice or walnut shells, laser ablation	Remove heavy scale, heavy oxidation, iron stains, mineral stains
4	Mechanical finishing	Can selectively remove areas. Can remove scale and patinas.
5	Laser ablation	Waste is limited and contained by the vacuum.

In laser ablation, the energy is absorbed into the oxide and it sublimates from the copper alloy surface. The energy effectively breaks the bond holding the oxide to the copper. The result is a chemically cleaned copper alloy (refer to Table 8.2).

ACHIEVING MECHANICAL CLEANLINESS

Achieving mechanical cleanliness involves restoring a surface that has been scratched or dented.

Because of the inherent softness of copper and some of the copper alloys, dents and scratches are a significant concern. When they occur, repair can be difficult. Copper is relatively soft and can be abraded by other materials. Table 8.3 lists a few materials and compares their relative hardness rating.

TABLE 8.3 Relative hardness of substances on the Moh scale.

Material	Moh hardness rating
Diamond	10
Glass	5.5
Steel	4
Copper oxide	3.5
Copper alloys	3
Aluminum	2.5
Talc	1

Scratches and Graffiti

Scratches and graffiti are an increasing issue with copper alloys used in proximity to human activity and contact. Unfortunately, many sculptures set in public spaces can attract people set on ignorant destruction. The soft metal and contrasting patina seem to attract the dissolute. Figure 8.26 shows examples of such wanton destruction.

FIGURE 8.26 Scratches through the protective layer and patina of three surfaces.

To restore scratched and marred surfaces to the artist's original intention can be a difficult procedure. First the coating and patina around the area must be removed down to the base metal. The process then involves grinding and polishing until the scratch is diminished sufficiently. The surface will be clean and smooth, with the base metal around the scratch exposed. The texture will be smooth and different from the surrounding metal. Initially you will need to match the texture of this surrounding metal, and this may require roughening with a small blasting tool; different media may need to be tested to arrive at a matching surface. A patina will also need to be developed to match the surrounding patina. This may not be possible, and the entire piece or segment of the sculpture may require stripping and repatination. Stripping may involve pressure blasting the surface with walnut shells or dry ice.

Thus, to achieve mechanical cleanliness some of the surface must be removed. For patinated or oxide surfaces, these need to be taken all the way down to bare metal. The scratch will first be sanded, then buffed to match the surrounding texture. Once this is achieved the patina or oxide can be applied.

Handling and Packaging

Copper alloy fabrications should be handled in such a manner as to not damage the surface or introduce substances that will initiate oxidation or lead to tarnish developing under the film if the surface is coated. Wear cotton gloves while handling the fabricated surfaces, store fabrications in dry, low-humidity places, and protect the copper alloy with protective wraps during handling and processing. Corrosion-inhibiting paper wraps that have benzotriazole-saturated layers can be used to individually wrap copper alloy fabrications or to line the containers or crates used to ship or store the parts.

Careful packaging of copper alloys is critical to prevent moisture from entering their crates (refer to Figure 7.22). Copper and its alloys need special care, and planning for and preventing damage while in transit is an important step. Fabricated copper alloy parts should be separated to keep them from rubbing on one another. The parts shown in Figure 8.27 are being shipped overseas in a container. Each plate has been fixed to a rigid wood frame and accurately marked for placement on the project. An anti-moisture dry pack of silica gel is placed in each crate to absorb condensation moisture during shipping and subsequent storage. The crates are also lined with a water-deterrent film to shed any moisture that might get on the crates and seep into the interior.

These precautions should be performed even if the copper alloys are coated with a protective layer of lacquer. Copper alloys have such an affinity to moisture that it is paramount to keep the fabricated work dry during storage and shipping. Granted, if the copper is downpipe or metal roofing the precautions can be reduced, but it is very important not to allow moisture to collect on a copper alloy surface during storage and shipping.

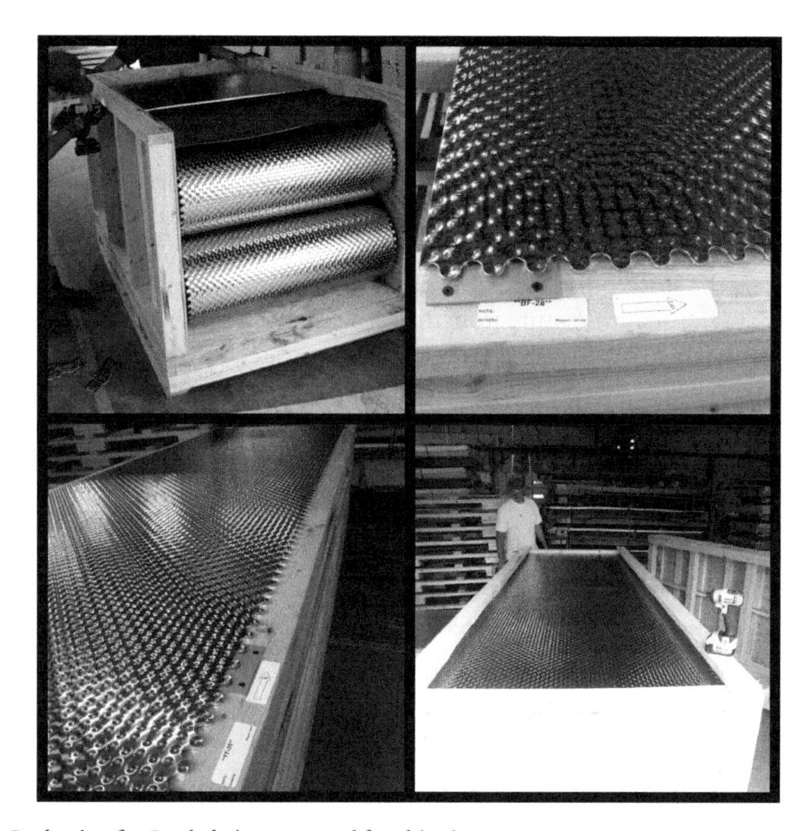

FIGURE 8.27 Packaging for Prada being prepared for shipping.

Maintenance Needs and Reasons

There are any number of maladies that can inflict themselves onto the surfaces of copper alloy statuary.. Coarse surfaces make cleaning, waxing, and polishing the surface difficult. On exterior pieces, water can infiltrate some porous regions and slowly drain out, concentrating into streaks or resisting wax and lacquer adherence.

Wrought surfaces—those copper alloy sculptures that are made from sheet, plate, or wire—have maintenance needs similar to cast surfaces. These surfaces can also come into contact with substances that can cause a chemical reaction with the copper or the alloying elements in the metal, with zinc being one of the more critical of these elements.

In copper alloys, the presence of moisture is necessary to start the deterioration of the surface. The presence of moisture is all but unavoidable. Condensation, the relative humidity of a space, and moisture in other materials make isolation impractical. One can slow its access to the surface, but it will eventually arrive there. Moisture gives other substances that may be on the surface the ability to react. Moisture has both hydrogen and oxygen, so the preliminary substances needed to create an acidic or alkali environment are present. Corrosion will now depend on the other salts that may be on the surface and the temperature and humidity.

Corrosion products transferred from nearby steel structures can stain a copper surface. These stains will show as a contrast to the patina on the copper alloy. They are difficult to remove without damaging the existing oxide or patina on the surface. Figure 8.28 shows several examples of steel corrosion particles on the surface of copper work.

Foundries work to eliminate moisture from the surface by heating the metal up prior to applying the patina or protective waxes. This will greatly reduce the presence of moisture on the surface of the casting. One issue, however, can be moisture entering the casting due to porosity. Some castings may contain pinholes and cracks that allow moisture to enter into the hollow interior. If there are steel structures or pins used to anchor the castings, these can corrode and the steel corrosion products leach out, staining the surface (Figure 8.29). These are difficult to remove without harming the patina on the bronze casting. If there is still a decent wax coating or lacquer coating, then the rust

FIGURE 8.28 Steel corrosion products on several surfaces.

FIGURE 8.29 (*Left, top right*) Rust stains on cast sculpture; (*top right*) before and (*bottom right*) after removal of the protective coating.

runoff will be on the coating and not the patina. Removing the wax or the lacquer will take off the rust with it.

Another issue that can develop given the porous nature of cast metal occurs when the foundry has left the core or portions of the core inside the casting. The core is the interior fill material that occupies the void in a bronze casting. The core material, called "luto," is made of refractory material such as baked plaster and silica molds that have been crushed into a fine powder. Some of this will stick to the interior side of a casting when heated and needs to be broken off when the casting is completed.

When the cast sections are removed from the mold, the interior surface usually has the refractory material removed mechanically. Depending on how a casting is made, removing all the core material can be difficult. When core material remains, it can absorb moisture that enters though gaps and pores in the metal and then leaches out, creating areas where the copper, patina, and the core moisture react to form blisters. Figure 8.30 shows significant blistering of the surface of a sculpture that had core material left in it. The water leaching out from the interior has ruined this surface.

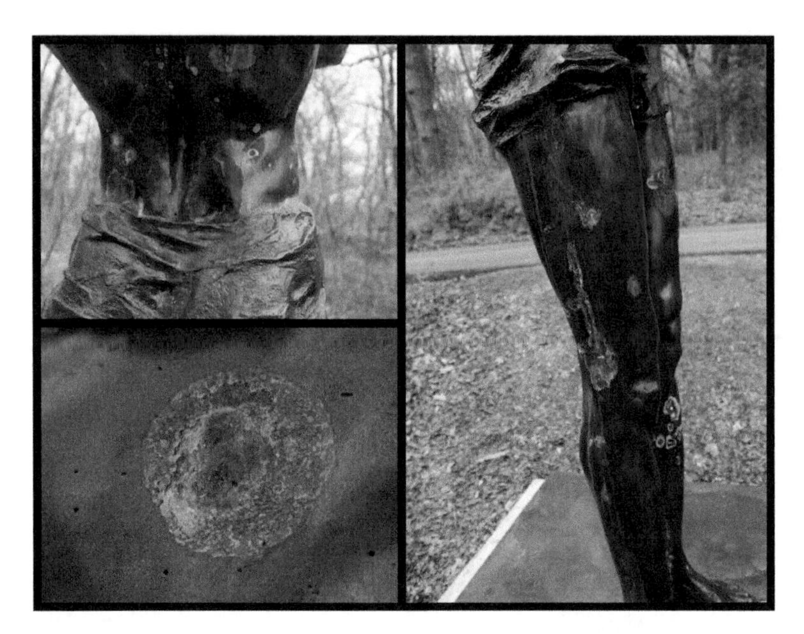

FIGURE 8.30 Blistering on the surface due to the core remaining inside the sculpture.

Such damage does not reveal itself initially. The sculpture in Figure 8.30 was set outdoors and it took a year for the damage to begin to appear. Once it did, however, it spread across the surface. This kind of issue requires moisture to manifest itself. If the piece had been placed inside, this would not have occurred—but in any case removing the core is standard practice for reputable foundries.

REPAIRING PATINAS

On the occasions where the patina has been damaged or needs to be restored, knowing what the metal constituents are or what the closest alloy is, is helpful. If any welding is needed you will want to know what alloy you are working with before you begin. If this information is not known, there are methods available to determine the metal constituents and the closest alloy type. One method commonly used by the recycling market to determine the various component materials being recycled is called X-ray fluorescence (XRF).

Determining the Alloy

Figure 8.31 shows a light fixture before restoration on the left and after restoration on the right. To determine what the manufacturer used to produce the piece many decades before, XRF readings were taken of the piece. The analysis showed that the alloy used in the casting was C85800.

FIGURE 8.31 Antique light fixture before (*left*) and after (*right*) restoration.

In this case we used a portable XRF device. XRF analysis involves bombarding a surface with gamma rays and reading the "signature" photon of the surface atoms. Essentially, if the energy of the high-energy gamma rays hitting the surface atoms of an object is greater than the ionization energy of the atoms, electrons in the inner shells can be dislodged. When this occurs, the atom will be unstable and electrons in the outer shells will fall into the inner shell to fill the gap left by the inner-shell electron. This releases energy specific to the differences of the two electron orbitals. This specificity is a signature of metal atoms and the XRF compares this released energy to a database.

Figure 8.32 shows two readings taken of the top portion of the fixture. The XRF device reads the energy and compares it to a database of energy levels to identify the precise atoms and their percentages to determine which alloys closely match.

From the readings it was apparent that the small leaf forms were cast from a leaded brass that closely matched the C34000 alloy. More likely was that the cast alloy would have been close to C85800, also known as leaded yellow brass. This was a common alloy used to cast small ornamental parts in die cast molds. The size, shape, and detail of these leaves point to this. The top spun lid showed as C26000 (Cartridge Brass) in the XRF. This is most likely the case, since C26000 is a common spun alloy of copper. Readings are comparisons to what is in the data base. Further inquiry is needed to determine what adjustments are needed to interpret the readings correctly.

When restoring the patina, isolating the area as much as possible is suggested. That is, limit the surface to be restored to as small a region as practical, otherwise you will be stripping the entire surface and repatinating. Sometimes this might be the only choice.

FIGURE 8.32 XRF readings of the top section of the light fixture.

Once you know the metals, obtain samples of similar alloys and test the patina formulas that when applied will develop a close match to the color.

To begin, remove the existing patina using light abrasion and 60% phosphoric acid.

Laser ablation will also work well. Blasting the large surface area and taking it back to bare copper alloy will be required.

Correct any surface defects by sanding and polishing to match adjacent regions. Immediately after the base metal surface is exposed, begin applying the patina.

Develop the new patina, preferably with heat. Slowly build the color back to closely match the surrounding surfaces by stippling, brushing, or spraying the patina onto the warm surface. You need to keep the surface temperature around 90–95 °C (200 °F). Rinse the area and compare it to the rest of the surface in different light. Once you are satisfied that a close match has been obtained, coat the surface with clear lacquer or wax it. Again, compare the area to the untouched areas to see if the color gloss and sheen are acceptable. Often rewaxing or relacquering the entire surface helps to blend the appearance.

PROTECTING THE SURFACES OF COPPER AND COPPER ALLOYS

In the art world, cast bronze sculpture is often protected from the environment with a thin layer of clear wax. The general technique is to gently heat the surface of the sculpture and brush on a coating of wax. The heating is believed to open the pores on the surface, which the wax enters as it melts.

The wax is applied in layers, and preceding applications of wax are allowed to dry before applying the next layer. Once applied the wax can be buffed with a clean cotton cloth to produce a smooth, continuous layer over the surface.

The heating of the surface also removes moisture that may reside in the pores or in recesses on the surface of the sculpture. Wax applications over moisture or oily residue will fail. Water or oil will prevent the wax from adhering to the surface of the copper alloy. Small remnants of moisture will create small pores through the wax, allowing the external environment to reach the copper alloy. Therefore, it is very important to have a clean, dry, oil-free surface before applying wax to the surface.

Waxes

Waxes are hydrophobic, meaning they will not dissolve in water and water tends to bead and shed on a wax-coated surface. Waxes are soluble in nonpolar solvents, such as mineral spirits. Wax coatings have good adhesion to metal surfaces and to oxides and patinas created on the surfaces. They do not adhere well to other wax coatings, though. One or two layers of wax are all that is needed. Beyond that, the surface texture gets gummy because of wax building up on the surface. A thin wax layer is more desirable than a thick wax layer. A buffed and polished wax surface will impart a clear, hard, elegant surface, adding depth and luster to the copper alloy.

When applying wax, the surface should be warmed. This aids in the flow of the wax deep into the metal surface. Allow the wax to dry for several hours before buffing the surface. Buffing the wax spreads the wax crystals out over the metal surface, forming a tight barrier of protection. Use clean, soft, cotton rags or buffing pads to work the wax into the surface and produce a smooth lustrous film. You can tell if the wax has not sufficiently dried if the surface has a gummy feel as it is rubbed.

Natural waxes are produced by plant and animals. These are long alkyl chains that contain alkanes, fatty acids, and alkyl esters. Beeswax is an example of an animal-based wax, while carnauba wax is an example of a plant-based wax. Beeswax is produced from the secretions of worker bees. A worker bee has glands on its sides that secrete the wax as the hive is created. Refined beeswax is white and noncrystalline. It is sensitive to temperature and humidity changes, and because of this its use as a protective wax on metals is limited. Additionally, beeswax can have a corrosive effect on copper alloys as it decays.

The carnauba wax is a general-purpose hard wax used for protecting sculpture as well as floors and automobile surfaces. It is used in numerous other products, from chewing gum to lipstick. Carnauba wax comes from *Copernicia prunifera*, a Brazilian palm tree that is known as the wax palm or carnauba palm. This hard wax is extracted from the underside of the palm leaf. It has a high melting temperature and often is mixed with other waxes to increase their melting temperatures. Carnauba waxes can be applied to produce hard, hydrophobic surfaces. When they degrade, however, these plant-based waxes can become acidic. For sculptures manufactured from copper alloy, this decrease in pH will usually not be a concern.

Another wax derived from plants is candelilla wax. This wax is made from the leaves of a small plant called the candelilla that grows in southern United States and northern Mexico. The wax is composed of hydrocarbons and is usually mixed with other waxes. It is yellow in color.

Synthetic waxes are produced during the petroleum refinement process. Two major types used in art and architecture as protective waxes for metal surfaces are paraffin waxes and microcrystalline waxes.

Paraffin waxes are large-crystalline waxes produced by a dewaxing process in the production of paraffin distillates. These waxes have low toxicity and good water-repellent characteristics, but the large crystalline structure makes them difficult to work into metal surfaces.

In common use as a protective coating for copper alloys are the microcrystalline waxes. Microcrystalline waxes are petroleum-based waxes with good flexibility and hardness. They are made up of fine particulates (thus the name) and will coat a surface evenly. The small crystals are long branching carbon chains that give these waxes good flexibility. Sometimes mixed with carnauba wax to create blends, microcrystalline waxes are considered to be more stable than carnauba waxes or animal waxes and will not become acidic as they degrade. These waxes have good water-repellant characteristics. A microcrystalline wax known as Renaissance Wax™ was developed by the British Museum in the mid-1900s. This is a common wax for protecting metal surfaces and will remain at a neutral pH as it degrades.

The harder carnauba wax is better suited for outdoor sculptures than the basic microcrystalline wax. A blend of the waxes gives best results outdoors.

Butchers wax and Bowling Alley wax are blends of microcrystalline wax and carnauba wax. There are other blends on the market as well, but the combination of microcrystalline and carnauba wax has shown good durability and performance.

Wax	Characteristic
Beeswax	Temperature sensitive
Carnauba wax	Hard and durable; hydrophobic; higher melting temperature
Candelilla wax	Hard; yellow in color; usually used in a blend with other waxes to facilitate hardening
Paraffin	Large-crystalline structure; difficult to apply; synthetic
Microcrystalline	Synthetic wax with good durability; lower melting temperature; high molecular weight; should be blended if used on exterior sculpture
Blend	Improved characteristics by combining waxes

The biggest drawback to wax protection of copper alloy sculpture and artifacts is the short life span of the compounds. Over time they will dry out and shrink. Degradation first appears as whitish or yellowing streaks on the surface. As further degradation occurs, the wax coatings crack and wear away from the surface, leaving the metal—or the metal oxide, in the case of a patina—exposed. You can expect a service life of two to five years when the wax is applied correctly. Temperature and humidity will determine the long-term performance of waxes, as will abrasion from human interaction or other contact.

When salt-laden moisture is able to pass under the wax coating and become trapped against the copper alloy sculpture, the potential for bronze disease can arise. Spots of pale green will appear as an indication that the wax coating has been breached and copper chloride compounds are forming.

Wax coatings can be more easily removed from the surface of the metal with the use of nonpolar solvents, such as naphtha-based solvents like mineral spirits. This is a distinct advantage over lacquers, which entail more rigorous polar solvents. To remove the wax, heat the surface of the metal up to around 90 °C (200 °F). Moisten a clean rag with mineral spirits and wipe the surface down, the old wax will loosen and come off onto the rag. Exercise caution when using open flames near mineral spirits.

Organic Coatings: Lacquers

Lacquer coatings were created specifically for copper alloy surfaces. They contain oxide inhibitors and ultraviolet radiation absorbers. The oxide inhibitors are specially formulated to combine with copper atoms on the surface, effectively changing the energy level and the drive to combine with oxygen.

There have been numerous tests using azoles, amines, and amino acids.[1]

In art and architecture, the most commonly prescribed lacquer, Incralac, is composed of benzotriazole in an acrylic resin known as B-44 and a solvent blend. Ultraviolet absorbers and a chelating agent to combat under-film corrosion are usually included. Chapter 4 describes the application of this coating.

Benzotriazole is the combination of an organic benzene molecule with a triazole molecule forming the organic molecule $C_6H_5N_3$. It is the nitrogen ion that appears to provide the molecule with the oxidation inhibitor for copper. The nitrogen ion has free electron pairs that bond with the copper ions on the surface. This effectively prevents the copper ion from bonding with oxygen. The mechanism is referred to as "chemiabsorption," in which the surface of the copper bonds with the benzotriazole molecule at the nitrogen site. Refer back to Figure 4.11 for an image of the molecule and where it is thought to bond with the copper atom at the surface. Further, the triazole molecule fills voids and shields the copper with a very thin layer across the surface.

There are other azoles that have shown promise in inhibiting oxidation on the surface of copper alloys. These involve the tetrazole molecule. This synthetic molecule also has nitrogen in the compound (CH_2N_4). The compound phenyltetrazole $C_7H_6N_4$ is also a good corrosion inhibitor and works in a fashion similar to benzotriazole.

There are two issues with acrylic lacquer coatings: under-film oxidation in the form of small dark spots and under-film oxidation along the edges in the form edge darkening. These issues only present themselves on copper alloys that are polished and intended to show natural copper alloy color tones. When darkened by statuary oxidation processes or patinated to develop rich color tones, this darkening may occur, but it will be concealed by the lack of contrast.

---

[1] M. M. Antonijevic and M.B. Petrovic, A. Macchia, M. P. Sammartino, and M. Laurenzi Tabasso, "Copper Corrosion Inhibitors. A Review," *International Journal of Electrochemical Science* 3 (2008): 1–28.

This type of oxidation is caused by improper preparation of the metal surface. Moisture left in the pores of the metal from rinse water or condensation that formed prior to the application of the lacquer will slowly darken the metal surface under the lacquer. The curing of solvents can also create this darkening. The solvents can react with ultraviolet radiation and cause oxidation reactions under the film. This problem can also arise if the chelating component of the lacquer has deteriorated or the coverage was insufficient.

It is important to use a good copper corrosion inhibiting lacquer with proper chelating and ultraviolet light inhibitors. It is extremely critical to start with a clean, dry surface.

When the lacquer fails, it must be removed from the surface. The oxide must be eliminated by treating the metal surface and this may entail restoration of the finish or patina on the copper alloy. This can be a costly exercise, in particular on large surfaces.

The best route is to prevent under-film darkening from occurring in the first place. This can be done by heating the metal to drive the moisture out before coating it with the clear lacquer. Follow with a wipe or rinse of 2% benzotriazole dissolved in alcohol. This establishes an oxide inhibitor across the surface so that subsequent handling and temporary exposure will not allow moisture to take hold. Be cautious about using flammable substances such as alcohol and lacquers near heat sources. Apply the first coating of Incralac, diluted to a 3:1 ratio of Incralac to xylene or toluene. Allow this to dry thoroughly. Apply a second coating, usually less dilute, and then a final layer of the lacquer. Allow the Incralac to dry thoroughly between each coating.

Degradation of Organic Coatings

The organic coatings used on metals are made of polymers. A polymer is a long-chain molecule made up of monomers and linked together by covalent bonds. As an example, polyethylene is a polymer of the monomer ethylene. These covalent bonds link the molecules together into chains.

In the acrylic lacquers used on metals the acrylic polymer has a photo stabilizer added to prevent photodegradation of the polymer chain. The benzotriazole helps to stabilize the acrylic against sunlight damage. So, it works both as an ultraviolet stabilizer for the acrylic and as an oxidation inhibitor for the copper.

When the acrylic coating fails, the failure begins first along the edges where the coating is usually thinnest. The coloration will appear darker than the rest of the surface. On patinated surfaces the darker color may be concealed, and eventually moisture and pollutants can change the color of the patina. But on natural polished or brushed surfaces and on light-to-medium statuary finishes there will be a darkening effect.

Inorganic Coatings

Inorganic coatings used as oxide inhibitors have shown promising results. These are considered alternatives to the organic lacquer coatings because they do not degrade as the organic coatings do. Organic coatings have a workable useful life of from 7 to 20 years, depending on quality of application, exposure and maintenance. Some have performed outdoors for longer periods.

Silanes and siloxanes polymers are inorganic coatings used as corrosion inhibitors. These are organosilicon molecules that are very hydrophobic. These silicon-based compounds are used as coatings to repel moisture. They are the basis for many anti-fingerprint coatings. These coatings form a thin layer over the copper alloy surface that is only a few atoms thick. The major drawback is they cannot be removed easily. They will swell when addressed with solvents, but solvents do not dissolve them. Siloxanes cure into cross-linked molecules that are highly chemical resistant. The only way to remove them is to break the silicon bonds. Acids and alkalis are ineffective. There are commercial silicon-bond "digesters" available. They are applied to the surface and allowed to set for a period of time as they break the bond with the copper alloy surface. But, as with the organic coatings, this is impractical, if not impossible, for large scale surfaces.

Oils

During the past century, many bronze sculpture and even copper architectural surfaces were coated with drying oils. Several of the more common were linseed oil, lemon oil, paraffin oil, and castor oil.

Linseed oil was perhaps the most favored of these oils. It once was used as a sizing in gold-leafing applications. Not in common use today, this oil has good hydrophobic qualities. Linseed oil is a natural hydrocarbon-based oil derived from flax. When it dries it leaves a hard, clear coating. It has been used as a protective coating for copper roofing and sculpture in the past. Linseed oil consists of unsaturated fatty acids, stearic acids, and palmic acids. It dries and eventually weathers off of exposed surfaces.

The other oils mentioned—lemon oil, castor oil and paraffin oil—also act as hydrophobic coatings on the metal surface. These oils are infrequently used today on metal art and architecture.

CLEANING THE COPPER SURFACE

A clean copper alloy surface is necessary to ensure good contact and coverage of applied inorganic or organic coatings. The creation of statuary and patina surfaces requires a clean metal surface. All coatings, whether applied or chemically induced, must link to the metal surface. Oils, excess moisture, and other substances on the copper alloy surface will interfere with that contact. Copper alloys can be cleaned and degreased effectively by using alkaline solutions, such as degreasers with sodium hydroxide or potassium hydroxide. These will remove oils, grease, and other organic emulsions that can be left on copper surfaces during manufacturing and handling processes.

When copper alloys are heat-treated with lubricants on the surface, they can become fixed into the metal and will require a robust treatment to remove them. Acid treatments consisting of a mixture of sulfuric and nitric acid work well in removing thick oxides. Mechanical means may also be needed to remove thicker oxides.

Light tarnish and oxide can be removed using milder acidic solutions, such as phosphoric acid and concentrated citric acids. These will remove cupric oxide but will not remove cuprous oxide.

The following is a list of various chemical solutions that have shown success in cleaning copper alloys:

- Ammonium citrate (5% solution)
- Citric acid (20% solution plus 4% thiourea)[2]
- Phosphoric acid (10–20% solution plus 1% thiourea)
- Ethylenediaminetetraacetic acid (EDTA) (4% solution)
- Potassium sodium tartrate (25% solution)
- Sodium hydroxide with glycerol (120 g/40 g per 1 l H_2O)
- Polymethacrylic acid (10–15% solution)

Copper alloys can be electrocleaned effectively in 2–5% sodium hydroxide solution.

Selective electrocleaning with 60% phosphoric acid solution is another effective way of removing oxides, dezincification deposits, and cuprous oxide formations from the surface of copper alloys.

REMOVING COPPER STAINS FROM OTHER SUBSTANCES

As copper and some of the copper alloys develop their natural patinas, often some of the compounds involved will dissolve into moisture from rain and condensation. If there are porous, light-colored substances below or in nearby proximity to copper surfaces, copper salts can redeposit into their pores by means of capillary action. Figure 8.33 shows images of a 100-year-old stoplight made from Monel. Monel is a nickel–copper alloy with excellent corrosion resistance once favored as a sheet-metal material. The copper gave it good formability. As this alloy weathers it turns from a gray nickel tone to a ruddy brown color. Monel has about 30–33% copper and it was the copper that left the stain on the limestone base of the light shown in the left-hand image.

These copper deposits can have a detrimental esthetic effect on appearance. Limestone, concrete, travertine, and other light-colored, porous materials can show these streaks and stains. The deposits can be tenacious and removing them can be very difficult. Steam, pressure washing, and detergents have little effect in the removal of the stains. Current approaches involve introducing a solution or paste that combines with or displaces the copper and allows it to be lifted out of the pores.

There are several commercial treatments that have proven results. These involve ammonia: usually ammonium carbonate made into a thick paste or poultice. Other substances are added to act as a surfactant or chelating agent and allow the copper salts to be displaced.

EDTA and Laponite® are often added to aid in the displacement. The poultice is allowed to remain on the stone surface for several hours and then rinsed off. Due to the heavy ammonia base, other surfaces as well as the workers should be protected. Powerful ammonia applications will require appropriate ventilation and eye, skin, and lung protection at a minimum.

[2]Thiourea is an organosulfur compound similar to urea.

FIGURE 8.33 Cleaning the copper stain from a limestone base, before (*left*) and after (*right*).

There are additional treatments that are showing promise in removal of these stains by less hazardous means.[3] One uses alanine, an amino acid. Amino acids are organic compounds; due to their makeup of both amino and carboxylic functions they can take on many forms. They are often used as release agents, detergents, and dyes, among other things. When accompanied with lesser concentration of ammonia, the alanine is found to be very effective in removing copper stains without presenting the risks associated with the handling and disposal of hazardous materials.

DETERIORATING PATINAS

There are occasions in which the patina on copper surfaces can deteriorate or fail. This can occur when other substances, such as iron or chemical sprays, have gotten on the patina surface. Muriatic acid, a dilute form of hydrochloric acid used to clean stone, brick, and concrete can cause an artificial patina to lift from the copper surface. Ammonia-based cleaners used on glass or ceramic will affect

[3] A. Macchia, M. P. Sammartino, and M. Laurenzi Tabasso, "A New Method to Remove Copper Corrosion Stains from Stone Surfaces," *Journal of Archeological Sciences* 38, no. 6 (June 2011): 1300–1307.

the patina's color and can cause the patina to lift from the underlying metal. Even water, if it is allowed to sit on the surface, can damage a patina.

Furthermore, if the patina was improperly applied or the surface of the copper alloy was not correctly prepared and cleaned, it can lead to the patina or oxide sluffing off of the surface or becoming streaked. To address these conditions, it may be necessary to remove all or part of the patina from the surface.

Attempts to touch up areas can be futile. For sculptures, isolated areas can be addressed. The geometry of the sculptural form may allow for conditional limits that can conceal the transition from the older patina to the new patina. Many in conservation address small patina blemishes by utilizing infilling techniques that use paint or colored waxes. This will work for a while, but eventually the surface will need to be addressed with a more in-depth restoration.

On architectural surfaces, it is ill advised to attempt applying fixes to a patina failing in place. The copper alloy surface may have a light oxide around where the patina has come away from the surface. Selective preparation of the surface will be difficult and matching an existing finish, in particular a shop-produced finish, will not be possible due to the variables involved.

Removing the patina from the entire surface and either allowing it to grow back over time (which will take decades) or replacing it with shop-prepared metal is recommended. Even if you decide to apply the patina on the surface in situ, then you must start with removing the oxide back to the bare metal surface.

APPENDIX A

Comparative Attributes of Metals Used in Art and Architecture

Various perspectives are presented in the following comparisons of the attributes of metals used in art and architecture.

The first is from the point of view of design and the ease of creating the form. The ease of shaping copper and aluminum make these metals the most versatile. Zinc is a ductile metal, but it has anisotropic properties and temperature constraints that will impact forming operations. Stainless steel and weathering steel are harder metals that will require more force to work, although stainless has good ductility for those that have the correct equipment. In the fully annealed state, titanium is ductile but difficult to machine, cast, and extrude. The Relative Attribute Value is a rough scale where 10 is the highest.

The tactile and surface color tones of these metals are described next in the following table. Copper is the densest of the architectural metals. Alloying of copper reduces this mass in most instances. Objects made of copper alloys have a "feel" of firmness to them. The color and appearance of the major architectural metals are all silver or gray, with the exception of copper alloys.

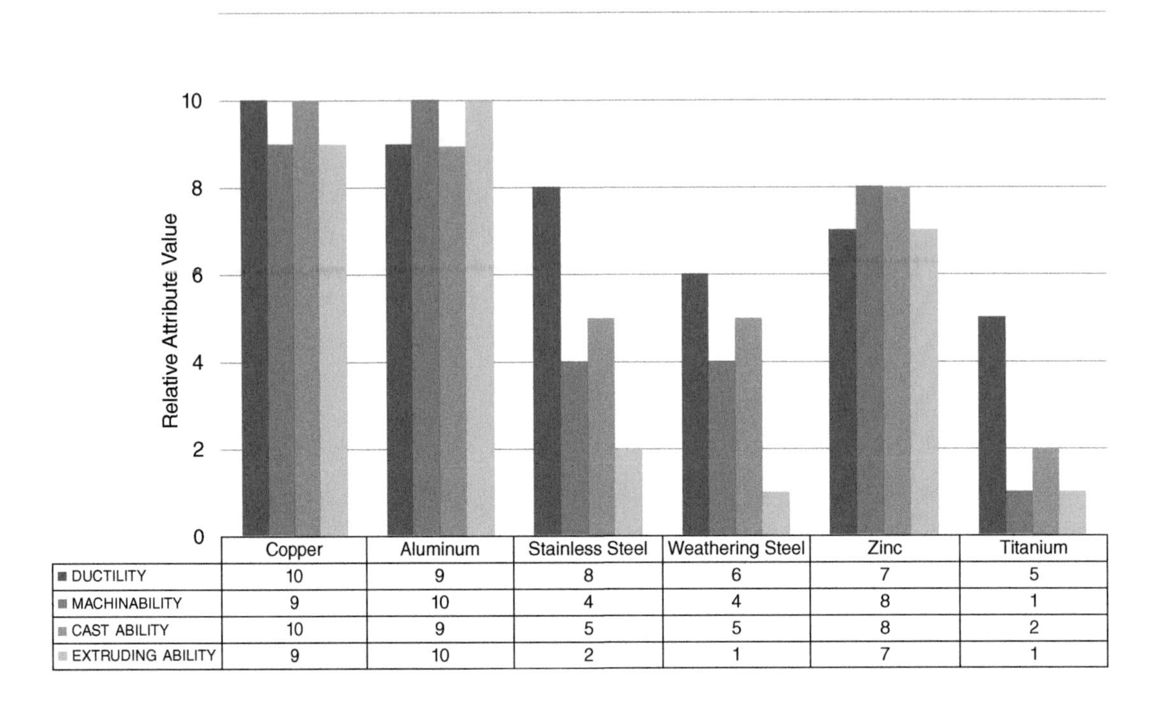

	Copper	Aluminum	Stainless Steel	Weathering Steel	Zinc	Titanium
DUCTILITY	10	9	8	6	7	5
MACHINABILITY	9	10	4	4	8	1
CAST ABILITY	10	9	5	5	8	2
EXTRUDING ABILITY	9	10	2	1	7	1

Attribute	Copper alloys	Aluminum	Stainless steel	Zinc	Weathering steel	Titanium
Mass	Densest of the architectural metals 8600 kg m^{-3}	Lease dense of the architectural metals 2560 kg m^{-3}	Dense 7600 kg m^{-3}	Dense 7135 kg m^{-3}	Dense 7850 kg m^{-3}	Low density 4500 kg m^{-3}
Natural color	From red to yellow to silver	White gray	Silver	Gray	Gray	Silver to golden silver
Change with exposure to normal environments	Will darken and develop patinas	Will darken slightly	Little noticeable change	Will darken; little change	Develops a orange oxide quickly	No change
Patina color	Yellow, red, green, blue green, black, brown	Multitude of colors from anodizing	Black, dark blue, bronze, gold, green red	Gray blue, variegated golden brown, dusty white	Deep reddish purple	Silver, yellows to magenta from anodizing
Reflectivity	Long wavelengths	90% visible light	60% visible light	Matte bluish gray	Matte, dark reddish purple	Matte gray, gold tint

The last graph compares current material costs for wrought forms of these metals and the ease of recycling them using available domestic processes. For example, titanium is recyclable but requires finding a company that will buy the scrap. Likewise, zinc is often painted on the reverse side to deter undersurface corrosion; the paint will reduce the recycled value but the metal can still be recycled at most centers that take it.

Aluminum scrap and waste collection operations are well established. Copper and copper alloys rarely make it to the waste heap, however—the difficulty lies in plastic-coated copper wire products. Such processing issues (and theft) place constraints on what the recycling market will pay for these metals.

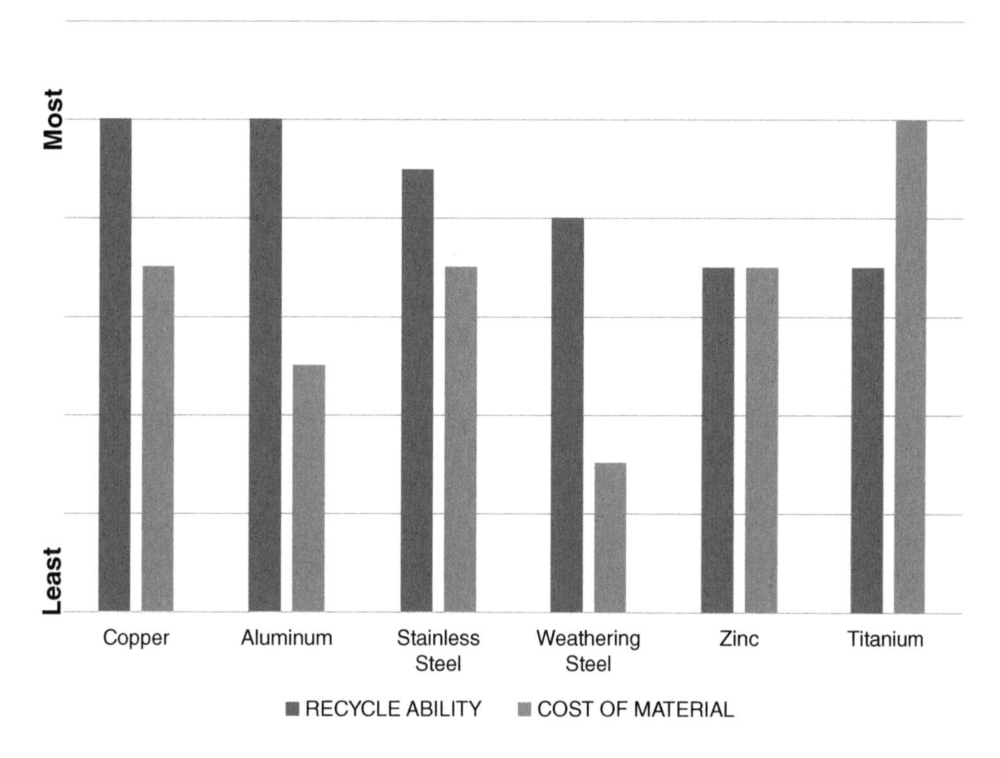

APPENDIX B

Hardware Finish Codes and Descriptions

A particular hardware finish is often specified for an architectural surface made from a copper alloy or alloys. It is important to establish a clear prototype that will serve as a target match for the supplier. Exact matching is very difficult and not practical, but in working with the manufacturer you can usually achieve some reasonable representation. Each manufacturer has variations in its processes for creating these finishes, and this needs to be understood by the design team and specifier.

Hardware Finish Codes and Descriptions

Description	ANSI code	US code	Close equivalent in wrought copper alloy, sheet, plate, rod, or tube
Bright brass, clear coated	605	US3	This corresponds to one of the mirror, or high-reflective polished yellow or gold brass alloys C26000, C27000, or C28000 with a clear coating applied.
Satin brass, clear coated	606	US4	This corresponds to an Angel Hair, glass bead, or hairline finish on brass alloys with a yellow or golden color, such as C26000, C27000, and C28000, with a clear coating applied.
Oxidized satin brass, oil rubbed	607	—	This corresponds to an Angel Hair or hairline finish on brass alloys with a yellow or golden color, such as C26000, C27000, and C28000, that have been oxidized to a medium statuary finish with a clear coating applied.

Description	ANSI code	US code	Close equivalent in wrought copper alloy, sheet, plate, rod, or tube
Oxidized satin brass, relieved clear coated	608	—	This corresponds to an Angel Hair or hairline finish on brass alloys with a yellow or golden color, such as C26000, C27000, and C28000, that have been oxidized to a medium statuary finish, then lightly rubbed to bring the color of the brass through and followed with a clear coating.
Satin brass, blackened, satin relieved, clear coated	609	US5	This corresponds to an Angel Hair or hairline finish on brass alloys with a yellow or golden color, such as C26000, C27000, and C28000, that have been oxidized to a medium statuary finish, then lightly rubbed to bring the color of the brass through along edges and followed with a clear coating.
Satin brass, blackened, bright relieved, clear coated	610	US7	This corresponds to an Angel Hair or hairline finish on brass alloys with a yellow or golden color, such as C26000, C27000, and C28000, that have been oxidized to a dark statuary finish, then lightly rubbed to bring the color of the brass through and followed with a clear coating.
Bright bronze, clear coated	611	US9	This corresponds to one of the mirror, or high-reflective polished copper or red brass alloys C21000 or C22000, or to the silicon–bronze cast alloy C87200, with a clear coating applied.
Satin bronze, clear coated	612	US10	This corresponds to one of the Angel Hair or satin finish red brass alloys C21000 or C22000, or the silicon–bronze cast alloy C87200, with a clear coating applied.
Dark oxidized, satin bronze, oil rubbed	613	US10B	This corresponds to one of the Angel Hair or satin finish red brass alloys C21000 or C22000 with a medium statuary finish, lightly polished, and with a clear coating applied. The intention is to make the surface looked aged. There is no oil used.
Oxidized satin bronze, relieved, clear coated	614	—	This corresponds to one of the Angel Hair or satin finish red brass alloys C21000 or C22000 with a light statuary finish and clear coating applied.
Oxidized satin bronze, relieved, waxed	615	—	This corresponds to one of the Angel Hair or satin finish red brass alloys C21000 or C22000 with a light statuary finish and waxed.

Description	ANSI code	US code	Close equivalent in wrought copper alloy, sheet, plate, rod, or tube
Satin bronze, blackened, satin relieved, clear coated	616	US11	This corresponds to one of the Angel Hair or satin finish brass alloys C21000, C22000, C26000, C27000, or C28000 with a dark statuary finish, followed by light sanding and clear coating applied.
Dark oxidized, satin bronze, bright relieved, clear coated	617	US13	This corresponds to one of the Angel Hair or satin finish brass alloys C21000, C22000, C26000, C27000 or C28000 with a dark statuary finish, followed by light sanding and clear coating applied.
Flat black coated	622	US19	This corresponds to one of the brass alloys C21000, C22000, C26000, C27000, or C28000 lightly bead blasted with a dark statuary finish and clear matte coating applied.
Light oxidized statuary bronze, clear coated	623	US20	This corresponds to one of the Angel Hair finish brass alloys C21000, C22000, C26000, C27000, or C28000 with a light statuary finish and clear matte coating applied.
Dark oxidized statuary bronze, clear coated	624	US20A	This corresponds to one of the Angel Hair finish brass alloys C21000, C22000, C26000, C27000, or C28000 with a dark statuary finish and clear coating applied.
Light oxidized satin bronze, bright relieved, clear coated	655	US13	This corresponds to one of the Angel Hair or satin finish red brass alloys C21000 or C22000 with a light statuary finish and clear coating applied.

APPENDIX C

Numbering Systems Used for Copper Alloys

Like languages, the numbering systems used for metals in different areas of the world have been adopted for reasons of national pride, legacy, colloquial, and more importantly today, digital efficiencies.

In Europe, the International Standards Organization (ISO) standard for designation of alloys is in common use. This is changing, however, as the CEN (European Committee for Standardization) system is being adopted.

For all metals and metal alloys, including copper, the ISO system lists the alloy components along with copper using an alphanumeric designation in which each primary element symbol is followed by its nominal percentage For obvious reasons this can become unwieldy and does not function well in the digital age. It is still used frequently by European suppliers, however.

The CEN designation, also a European standard, utilizes a more efficient alphanumeric system in which the first letter is a *C* for copper, followed by a second letter designating a *W* for wrought, a *B* for bar, a *C* for cast, and an *M* for master alloys. Three numbers representing the various copper groups by major alloying element follow the two letters.The Unified Numbering System (UNS) developed by the National Institute of Standards and Technology is similar to an old, three-digit numbering system created by the Copper Development Association. In North America it is more understood and frankly clearer and more straightforward than the ISO or CEN systems, at least in the context of alloys used for art and architecture.

The following table compares several common alloys. I will leave it to the reader to decide which system is easier to understand—I've made my choice.

ISO designation	CEN designation	UNS designation
CuZn5	CW500L	C21000
CuZn10	CW501L	C22000
CuZn15	CW502L	C23000
CuZn20	CW503L	C24000
CuZn30	CW505L	C26000
CuZn40	CW509L	C28000
CuZn39Pb2	CW612N	C37700
CuZn38Sn1As	CW717R	C46400
Cu 58	CW004A	C11000

Further Reading

ASM International (2001). *ASM Specialty Handbook. Copper and Copper Alloys*. ASM International.

Chawla, S.L. and Gupta, R.K. (1993). *Materials Selection for Corrosion Control*. ASM International.

Copper Development Association (1985). *Standards Handbook-Part 2-Alloy Data*, 8e. CDA.

Copper Development Association (1991). *Standards Handbook-Part 5-Sources-Part 6-Specifications Cross Index*, 5e. CDA.

Grilli, E., Notarcola, S., Blanco, F.S. et al. (2002). *Unknown Nobleness*. SMI Group.

Hayden, R.S. and Despont, T.W. (1986). *Restoring the Statue of Liberty*. McGraw-Hill.

Hurst, S. (2005). *Bronze Sculpture Casting & Patination: Mud Fire Metal*. Schiffer Publishing.

Kipper, P. (1995). *Patinas for Silicon Bronze*. Path Publications.

Lagos, G.E., Warner, A.E., and Sanchez, M. (2003). *Health, Environment and Sustainable Development*, vol. II. Canadian Institute of Mining, Metallurgy and Petroleum.

Pinn, K. (1999). *Paktong: The Chinese Alloy in Europe 1680–1820*. Antique Collectors Club.

Revie, R.W. and Uhlig, H. (2008). *Corrosion and Corrosion Control: An Introduction to Corrosion Science and Engineering*, 4e. Wiley.

St. John, J. (1984). *Noble Metals*, Planet Earth series. Time Life Books.

Copper and Brass Extended Uses Council (1927). *Copper in Architecture*. Copper and Brass Extended Uses Council.

Edge, M.S. (1990). *The Art of Patinas for Bronze*. Springfield, OR: Artesia Press.

Hughes, R. and Rowe, M. (1991). *The Colouring, Bronzing and Patination of Metals*. Watson-Guptill Publications.

MetalReference (2002). The 40 copper alloys [Copper alloys design guide]. https://www.metalreference.com/Forms_Copper_Alloy.html (accessed October 26, 2019).

Runfola, M. (2014). *Patina: 300+ Coloration Effects for Jewelers & Metalsmiths*. Interweave Press.

Scott, D. (2002). *Copper & Bronze in Art: Corrosion, Colorants*. Conservation: Getty Trust Publications.

Uhlig, H. (1948). *Corrosion Handbook*. Wiley.

Young, R. (2015). *Contemporary Patination for Bronze Brass and Copper*, 2e. Sculpt Nouveau.

Index

Note: Page numbers followed by *f* and *t* refer to figures and tables, respectively.

Printed and bound by CPI Group (UK) Ltd, Croydon, CR0 4YY
09/06/2025
14685920-0001